Discourse, Technology and Change

Also available from Continuum:

Using Corpora in Discourse Analysis
Paul Baker

Classroom Discourse Analysis: A Functional Perspective
Frances Christie

Historical Discourse: The Language of Time, Cause and Evaluation
Caroline Coffin

Language, Education and Discourse: Functional Approaches
Edited by Joseph Foley

Bridging Discourses in the ESL Classroom: Teachers, Students and Researchers
Pauline Gibbons

Scientific Discourse: Multiliteracy in the Classroom
David Ian Hanauer

Metadiscourse: Exploring Interaction in Writing
Ken Hyland

Expository Discourse
Beverly Lewin, Jonathan Fine and Lynne Young

Mathematical Discourse: Language, Symbolism and Visual Images
Kay O'Halloran

Multimodal Discourse Analysis: Systemic Functional Perspectives
Edited by Kay O'Halloran

Discourse Analysis: An Introduction
Brian Paltridge

Discourse, Technology and Change

Brenton Faber

continuum

Continuum
The Tower Building
11 York Road
London SE1 7NX

80 Maiden Lane
Suite 704
New York
NY 10038

Acknowledgements

A version of Chapter 1 was first published in Faber, B. (2007) 'Writing social change', in C. Bazerman (ed.), *Handbook of Writing Research.* Mahwah NJ: Lawrence Erlbaum Associates. Reprinted with permission.

Parts of Chapter 2 first appeared in Faber, B. (2003) 'Creating rhetorical stability in corporate university discourse: Discourse technologies and change', *Written Communication* 20(4): 391–425. Reprinted with permission of Sage Publications.

British Library Cataloguing-in-Publication Data
A catalogue record for this book is available from the British Library.

ISBN: 0-8264-94846
978-08264-94849

Library of Congress Cataloging-in-Publication Data
Faber, Brenton D., 1968-
Discourse, technology, and change / Brenton Faber.
p. cm.
Includes bibliographical references.
ISBN-13: 978-0-8264-9484-9 (hardcover)
ISBN-10: 0-8264-9484-6 (hardcover)
1. Discourse analysis. 2. Technological innovations. 3. Social change. 4. Written communication. I. Title.

P302.865.F33 2007
410'.41 - - dc22

2006038534

Typeset by YHT Ltd, London
Printed and bound in Great Britain by MPG Books Ltd, Bodmin, Cornwall

For Rebecca

Contents

Preface

Most people in modern workplaces have suffered through an organization-wide software implementation. In 2002, a small, rural technical university confidently launched a campaign to install WorkWare on all faculty, staff and student desktops. 'WorkWare', as the implementation documents boasted, 'is much more than just a new e-mail system. In addition to e-mail and messaging, WorkWare will enable integrated calendaring, document management & workflow, instant messaging and collaboration software, and an electronic workspace.' That is, until the first version of WorkWare arrived on campus as vapourware, an unfinished version of the software that quickly placed in crisis the implementation process, the corporate/university partnership that initiated the project, and the careers of the IT workers responsible for directing this change. What became remarkable about this campaign, however, was how a failed technological change sparked a significant social and cultural change as the IT workers responsible for installing WorkWare emerged from the implementation with new, empowered social roles within and across the larger organization.

Through a detailed examination of the texts written to convince users to adopt the software, even after the initial installations were less than optimal, this project demonstrates the intricate connections between discourse, technology and social change. This project initiates from Michel Foucault's challenge to 'establish the system of transformations that constitute change' (1972, p. 173). These transformations are examined as discursive and material forms/events and then articulated back against the social network in which they occurred. The result is an attempt to trace social change from disruption and instantiation, through resistance and various forms of contest, to some form of temporary stability. Thus, my analysis connects material events occurring on the ground to simultaneously generated, specific discursive features of the implementation texts. Chapters before and after the more granular discursive study address discourse and social change, the technologization of everyday life, the aggregation of discourse and technology, and the problem of why apparently offensive discourses still seem to be effective forms of social change. The conclusion offers the concept of a textnology as a unique and recent form/event of change that differentiates change as an active agent rather than a reaction to a pause or suspension in narrative.

I am borrowing the concept of a form/event from Paul Rabinow. He used the conjoined term to highlight the dual process in which events result from discursive and nondiscursive activities, make available new temporalities, 'and make new assemblages visible and describable' (1999, pp. 12–13). In other words, events create new forms which, in turn, make possible new events – an iterative and stochastic process of social change. As this book project demonstrates, this sort of iterative process is reflective of the dynamic networks that produce and enable change. In the book's conclusion, I consider these networks as stochastic systems and propose several methods for examining and assessing the social work they accomplish. The book's project is also ethical in that I wish to examine the value-work accomplished by the aggregation of discourse, technology and change and the ways this aggregation enables or resists projects of individual choice and accountability during the process of change. The book argues that discourse studies can examine and assess 'the ethics of change' by which I mean the social action instantiated in the ways agents choose to initiate, manage and stabilize change.

The story of WorkWare is situated within the field of discourse studies. Specifically, the project contributes to our understanding of the interconnection of language, technology and social change. I argue that discourse is a crucial component of change – that change is co-constituted by discursive and material actions – and as such can provide a useful analytical and empirical tool for assessing and examining change. The book has been developed to bring together audiences in discourse studies, rhetoric, organization studies, and management and technology studies. The book will also find an audience among technology practitioners and researchers in technical communication, professional communication, public relations, technology management and information technology.

This project has evolved through various iterations in presentations, early publications and informative discussions. These forms/events provided me with the selfish opportunity to test approaches, findings and speculations from what has been a unique data set. In each case, I came away with a renewed perspective of the project which influenced and altered my treatment of it in this book. I would like to thank audiences at the Texas Tech University Graduate English Society 2004 Conference, faculty and graduate students in the Department of Language, Literature, and Communication at Rensselaer Polytechnic Institute, and participants at SIGDOC 2003 in San Francisco. I would like to thank and acknowledge Mark Zachry and Charlotte Thralls who graciously provided me the opportunity to publish some initial findings and speculations as a contribution to their book *Communicative Practices in Workplaces and the Professions: Perspectives on the Regulation of Discourse and Organizations* (Amityville, NY: Baywood). The full citation is Faber, B. (2007) 'Discourse and Regulation: Critical text analysis and workplace studies', in M. Zachry and C. Thralls (eds), *Communicative Practices in Workplaces and the Professions: Perspectives on the Regulation of Discourse and Organizations.* Amityville, NY: Baywood.

I would also like to thank Charles Bazerman for his comments and assistance with Chapter 1, sections of which have appeared in his *Handbook of Writing Research* as Faber, B. (2007) 'Writing social change', in C. Bazerman (ed.), *Handbook of Writing Research.* Mahwah, NJ: Lawrence Erlbaum Associates. I gratefully acknowledge Lawrence Erlbaum and Associates for reprinting that chapter here. Thanks to Sage publications for permission to republish a revised version of pages 394–7 from Faber, B. (2003) 'Creating rhetorical stability in corporate university discourse: Discourse technologies and change', *Written Communication 20*(4): 391–425. I would like to thank the Alfred P. Sloan foundation for their financial support of my research into the phenomenon of the corporate university and its influence on academic campuses.

Special thanks to Rebecca Sutcliffe, Clay Spinuzzi, Johndan Johnson-Eilola, Geoff Sauer and Rosemary Brenan for their advice on previous versions of the manuscript. I would also like to acknowledge the exceptionally capable work of Jennifer Lovel and Adam Green at Continuum. Finally, I would like to acknowledge and thank Paul Rabinow whose concept of the 'anthropological contemporary' (to use shorthand) provided me with a model and perspective for interpreting and reporting my data. I recommend to interested readers his *French DNA: Trouble in Purgatory* (Chicago: University of Chicago Press, 1999).

Special note should be made to recognize and thank the participants in this study, Martin, Abigail and Kendra, who, according to the terms of my agreement with them, cannot be identified. Their willingness to support this project, to provide honest and open access to their notes, records, thoughts and recollections, and to work through this project with me, ultimately made this book possible. This fact alone must mediate their portrayal in the pages that follow. They found themselves in what came to be an extremely difficult situation and yet they continued to willingly and openly discuss their actions and ambitions throughout the process. For their honesty and openness I am deeply grateful.

1 Writing research and social change

> Since Kant, the role of philosophy has been to prevent reason going beyond the limits of what is given in experience; but from the same moment – that is, from the development of modern states and political management of society – the role of philosophy has also been to keep watch over the excessive powers of political rationality – which is rather a promising life expectancy.
>
> (Michel Foucault 1988, p. 58)

> The machine defends itself by devaluing analysis – or prose anyway.
>
> (Andrei Codrescu 2005)

In the mid-1980s a renewed presence emerged on college campuses in the United States. Emboldened by an emerging pragmatist academic culture and scornful of the perceived excesses of high theory in which campuses apparently were mired, corporations saw great opportunity in forging partnerships and buying influence within the college curriculum. Of course, corporate interest in colleges and universities is nothing new.[1] Yet, the form of influence developed at this time claimed a much more direct and immediate return on investment than previous models authorized. Whereas earlier corporate–government–university relationships were content to focus on technology transfer, collaborative research projects, various methods for attracting and recruiting students as future employees, and corporate benefactors for capital projects, this new model asserted that the fundamental culture of corporate capitalism could better organize and manage the business of academic teaching, research and student preparation. Advocates of what became known as 'corporate universities' claimed that for-profit companies were better suited for training employees for the jobs of the future. Not only would corporations make college campuses more relevant, they could also (finally) turn education into a profitable industry.

A key feature of modernity has been the dominant role written communication has come to play in both constituting and reflecting social practices (Chouliaraki and Fairclough 1999, p. 4). This dual function of discourse (constituting and reflecting) is especially apparent in the expansion of capitalist practices into university networks. The expansion of market interests into education was largely a discursive act, enabled through an aggregation of discourses which Gee, Hull and Lankshear called 'fast capitalist texts' (Gee *et al.* 1996, p. 25). 'Fast capitalist texts' are texts promoting global capitalism, the predominance of multinational corporations, and an embrace of the free market in all sectors of life, especially areas like education, which had remained independent from the organizing forces of the free market (Gee *et al.* 1996, pp. 26–31). As Gee *et al.* wrote, 'the fast capitalist version of the new capitalism is ... profoundly imperialistic, seeking to take over practices and social identities that are (or were) the terrain of other Discourses connected to churches, communities, universities, and governments' (Gee *et al.* 1996, p. 26).

What was significant about these new texts, like the relationships they attempted to renew between universities and the free market, was that they emerged from and were concerned with creating a very specific kind of social vision. Gee *et al.* called this motivation 'enactive texts' (Gee *et al.* 1996, p. 33), meaning visionary texts that attempt to create a vision of a new world which includes new values, social purposes and practices, and new social identities that efface older and existing values, relations and identities (Gee *et al.* 1996, p. 33).

Staged against this backdrop of the corporate university movement and emerging from a specific encounter with an enactive text at a rural technological school on the east coast of the United States,[2] this book is about the specific ways written communication constitutes and reflects social change. In social communities such as groups, institutions and communities of work, through the dual actions of constituting and reflecting, writing enables organization, maintenance and change. This book examines the function of written discourse across the temporal moments of change, the dynamic between written discourse and social context during change, and more pragmatically, it investigates how writing enables the initiation, management and stabilization of social and cultural change.

This chapter introduces and orients the contextual and methodological background and approach in which my project takes place. First, I will provide an overview of the current research and theory on writing and cultural change in social systems, primarily organizations. This research has been instrumental in helping scholars better understand and empirically document the function of written texts in networks of social change. I will explain that change occurs as a transitional process in which advocates will initially destabilize a current network, introduce new elements into that network, and then restabilize that network with new meanings and practices.

Second, this chapter introduces research that has attempted to better examine and understand the motivation or, in other words, the rationale for change. To pursue these questions, researchers have turned to the cultural-rhetorical concepts of image and identity, theorizing that rhetorical perception can play a central role in the reasons why social groups will enact processes of social and cultural change.[3] This chapter presents narratives as key organizational genres that functionally allow agents to challenge and replace a social group's image and identity. This more 'macro' discussion is followed by an outline of more 'micro' discourse-based research that has attempted to characterize the process of change at the linguistic level. This work has highlighted specific syntactic and pragmatic discursive features that appear to help texts create causal connections between old and new social networks. Throughout these sections of the chapter, I draw on the case of the marketization of university education for examples to demonstrate various theoretical and pragmatic concepts. To foreshadow a bit, Chapter 2 continues to present this narrative but in a situated account of how the corporate university emerged at a small American college, the scene in which this book takes place.

This chapter then concludes with a more theoretical discussion about approaches to integrating macro and micro approaches to text analysis, especially in the study of social change. A key aspect of this project for me has been to find and examine ways the largely disparate fields of rhetoric and discourse studies can together inform our understandings of texts, their networks, and how these respond to and constitute change. The chapter's concluding discussion attempts to bring together these textual practices in a way that enables researchers to document and examine the dynamic between textual activities and material conditions at multiple levels and purposes.[4]

Writing constitutes and reflects social organization and change

In their work on the origins and forms of writing, Schmandt-Besserat and Erard (2007) described a 'materialist' orientation toward writing which emphasizes the social activity that is enabled by writing. Citing Latour's (1990) studies of the function of writing within scientific communities, Schmandt-Besserat and Erard claimed that written representations 'succeed because they enable their users to do more' (Schmandt-Besserat and Erard 2007, p. 24). Thus, in areas of science (Lyotard 1979), cultural production (Jameson 1991) and in the emergence and cultural dominance of organizational and professional life (Burrell 1988; Foucault 1973, 1979a), written communication has functioned as a constitutive form of legitimacy and social power (Ong 1988) as it has enabled users to accomplish significant activities. The constitutive power and utility of written communication has been especially prominent in the creation and regulation of modern organizations and the varied and conflicting identities of those who work within these organizations (see, e.g., Whyte 1956; Mills 1951; Zachry and Thralls, 2007).

Although change is often associated with external, material conditions such as economic factors, environmental conditions or technological innovations, studies of the intersection of writing and social change have shown social change to be a multiformed process involving a complex interplay of discursive, material and social conditions. Legally, but also socially and culturally, modern organizations and professions are the products of written communication. Mumby and Clair claimed 'organizations exist only in so far as their members create them through discourse' (Mumby and Clair 1997, p. 181). This discourse functions 'simultaneously as both an expression and a creation of organizational structure' (Mumby and Clair 1997, p. 181). Similarly, Alvesson and Kärreman pointed out that throughout the social sciences, researchers have come to understand that 'societies, social institutions, identities, and even cultures may be viewed as discursively constructed ensembles of texts' (Alvesson and Kärreman 2000, p. 137).

As researchers have come to understand social groups and organizations as products of written communication, further studies have shown that social and cultural changes within such organizations are also combinations of social and textual activities. For example, researchers have documented the textual and rhetorical activities that have enabled conceptual changes in scientific fields (Bazerman 1988; Hull 1988; Myers 1990; Vande Kopple 1998, 2002; see Faber 2006 for discussion). David Hull went so far as to find that despite the process and chronology of scientific discovery, the written rationale for conceptual change is typically (re)ordered to assert causality and consistency with existing practices and research within the same field (Hull 1998, p. 116). It is important to note here that Hull's finding does not suggest impropriety or falsified data. Instead, this process is akin to the tension Deleuze and Guattari described between planes of organization and consistency (Deleuze and Guattari 1987, p. 270), or to the sort of retrospective regroupings by which Foucault (1972) characterized discipline formation and promotion. Knowledge is built in often chaotic and stochastic relations and formations. The interpretation of these formations re-aggregates the material conditions in stable, coherent accounts, creating this necessary tension.

(De)stabilization: change as transitional social-textual process

Paul Feyerabend, writing about changes in knowledge within academic groups, argued that change in disciplinary communities occurs primarily through a process of argumentation (Feyerabend 1991, p. 18; for discussion see Faber 2006). Using the term *transition*, Feyerabend argues that a successful transitional argument must coherently link from the existing to the new, 'protect' the new system from 'misinterpretation', and convince (condition) adopters that the new is vital and important (Feyerabend 1991, p. 18). This model emphasizes the creation of the new from within existing and disputed networks. In other words, this concept of *transitional change* occurs in increments or linked steps as prior existent knowledge is disrupted and eventually displaced by small additions that ultimately build into new formations.

Feyerabend's concept of transitional change was expanded by Bazerman

(1999) in his work on Thomas Edison's discovery and introduction of electric lighting technology (for a comparison of Feyerabend's and Bazerman's models of change, see Vanburen Wilkes 2002). Unlike Feyerabend's concept, Bazerman's examination was overtly rhetorical. He argued that both the technological discovery and the eventual widespread acceptance of electric lighting were dependent on Edison creating specific yet suitably varied social conditions that would welcome this new technology. Edison's project was equally rhetorical (discursive) as it was technical. It was also strategic in that genres, like patent applications, necessitated a written discourse that was strategically and generically different from other key written products (newspaper reports, consumer guides, advertisements).

Bazerman showed that Edison's work was incremental and transitional in the sense that it was a competitive enterprise that negotiated within and among others' projects, ideas and discourses. Bazerman introduced a useful vocabulary to describe this process whereby Edison's technology, in concert with his various discourses, introduced significant social and cultural changes. Bazerman used the metaphor of *stability*, noting that written documents will create 'stabilized representations' which articulate the final transition within a changing social network (Bazerman 1999, p. 308). It is important to note that the concept of stability also assumes that initial discursive activities will *destabilize* current networks. Thus, the 'coherent link' Feyerabend described often occurs as a subversive displacement. Written communication can provide a coherent link between old and new, but such a link can be used to displace and destabilize older (current) concepts and practices. This destabilization creates the conditions necessary for change.

Once a social network has become destabilized, writing will help to introduce emergent and competing alternatives (representations) and thereby introduce and stabilize the emerging system. In such a network, written communication can become highly strategic, controversial and negotiated at various levels as agents pursue competing and diverse representations. In this transitioning network, the emergent system requires protection (as Feyerabend suggested) in order to ensure that older practices do not re-emerge/restabilize. In addition, advocates of various alternatives will use writing (and other discourses) to promote and ultimately attempt to stabilize, their own systems.

As an enabling tool of social change, written communication makes possible the stabilization of new practices, and assists in their ongoing maintenance by allowing advocates to position and inscribe various representations of the new within the transitioning culture. To use Bazerman's terms, proponents create 'stabilized representations' (Bazerman 1999, p. 308) that compete with and eventually displace accepted concepts, meanings and interpretations (Faber 2003, p. 393). As older meanings are displaced, new meanings will congeal into what Bazerman called a 'unitary social fact' (Bazerman 1999, p. 144). Bazerman wrote that the new system aggregates together 'its own core of meanings, which draw the various other relevant discursive systems together in a new configuration' (Bazerman 1999, p. 144). In other words, the various competing discourses eventually aggregate together into a new model, which is represented as a core concept. This core concept then provides the basis for interpreting and perceiving other discourses and integrating them within the new system.

Example of destabilizing discourses: commodifying higher education

An example of this process can be found in the case of the marketization of higher education in North America and the United Kingdom. In the late twentieth century, the academic culture of university education in the US and UK was vigorously contested, and some would argue successfully displaced, by discourses of business and capitalism (for discussion see Lemke 1995, pp. 58–79; Fairclough 1992, pp. 208–9; Connell and Galasiński 1998; White 2000; Faber 2003). As noted above, Gee, Hull and Lankshear (1996) have described this encroachment of capitalist values into education as the activities of 'new capitalism' whose advocates do not limit market-based practices to business transactions but view the totality of human life through the lens of the market. The new capitalism is not based on material goods or exchange, but on values. Gee *et al.* wrote that the values of the new capitalism are 'profoundly imperialistic, seeking to take over practices and social identities that are (or were) the terrain of other Discourses connected to churches, communities, universities, and governments' (Gee *et al.* 1996, p. 26). In education, new capitalist discourses have aggregated together the core concept of 'educational market' and have used this concept to articulate new systematic roles that restructure other discourses of the university under this core concept. In this aggregation, students become customers, faculty –> retailers, research –> commodity, administration –> management, facilities –> resources. Fairclough showed that this market-based educational discourse has imposed a discourse of commodities onto learning, writing that the discourse:

> is dominated by a vocabulary of skills, including not only the word 'skill,' and related words like 'competence,' but a whole new wording of the processes of learning and teaching based upon concepts of skill, skill training, use of skills, transfer of skills, and so forth.
>
> (Fairclough 1992, p. 209)

Fairclough's analysis showed that this vocabulary of skills emerged from within the core concept (educational market) with new meanings and assumptions that carry specific normative and objective criteria that differentiate the commodified 'skill' from the liberal humanist notion of skill (individual ability). Fairclough has shown that this process often is accomplished through 'co-hyponyms' (Fairclough 2004, p. 111) in which certain words (skills/knowledge; training/education) are made equivalent under the core concept. In the perspective formed by the core concept *educational market*, skills/knowledge are gained from 'normalized [education/training] procedures' and they are '[taught/transferred] across contexts, occasions, and users' (Fairclough 2004, p. 209). As the new discourse stabilizes transitional terms (clients, retailers, commodities, management, resources), other, more radical terms (standardization, efficiency, cost containment) can be introduced legitimately.

Narratives: genres of image and identity construction

Students of classical rhetoric and sophistic education (Rankin 1983; Kennedy 1991; Jarratt 1992) will see concepts and assumptions from these traditions embedded within this model of a competitive and constitutive discursive process of change. Inasmuch as rhetorical debate and sophistic perspectives played key roles in forming, maintaining and changing the Greek city-state, these same principles continue to enhance our understanding of the functional role of written communication in the networks of modern organizational change. But, to better examine and understand the motivation or the rationale for change, researchers have turned to the cultural-rhetorical concepts of image and narrative, theorizing that rhetorical perception can play a central role in the reasons why social groups will go through processes of social and cultural change.

Using the metaphor of a mirror, organization studies scholars Dutton and Dukerich (1981) theorized that organizations present specific images to their members and to external audiences to build and reflect specific perceptions of the organization and its products and services. These perceptions enable employees and other organizational stakeholders (customers, suppliers) to define and identify themselves as part of the organization (Dutton *et al.* 1994). Dutton and Dukerich argued that image integrates how members identify with the organization and the extent to which they perceive that the organization shares common values. As Hardy, Lawrence and Phillips wrote, this image-driven concept of identity is important, 'not only because it affects others' valuations, but because it affects action: identity has been found to be a key factor in influencing whether issues are noticed, considered legitimate and important and, hence, acted on by different organizational members' (Hardy *et al.* 1998, p. 70).

An organizational image provides a way for an organization's members to identify with the organization and to assess their level of participation with various organizations. A similar concept, organizational narrative, has also been developed by rhetorical scholarship to identify a parallel process of identity construction within organizations (Boyce 1996). Mumby (1987, 1993) showed that organizational narratives are key to constructing social networks within organizations. He argued that narratives are important devices for meaning-making within organizations as they enable people to understand and explain their complex and often contradictory roles.

Narratives are forms of social reality, describing what Robert Heath simply called 'why we do things we do' (Heath 1994, p. 60). As stories of daily life, organizational history, historical events, or even rumours and gossip, the truth of the narrative is not found in its factual status but in the interpretation and the meaning it has for its teller and its audience (Gabriel 1998). Thus, like image, narrative also helps to define and reflect an organizational 'self' (Perkins and Blyler 1999, pp. 5–6). These stories present an organization's identity, its values, and the way it appears and is interpreted by various groups within and outside of its stakeholder community. Gee *et al.* described this function of narrative as 'enactive' since these texts are written (often by business leaders) to enact or 'call into being' a specific vision of a new organization (Gee *et al.* 1996, p. 33).

The discussion above reviewed various writing researchers who have posited that change occurs when discourses become destabilized. Destabilization occurs when an organization's narratives and images no longer support or construct a viable identity. Narratives and images will ideally work together to construct an organization's identity – providing complementary discourses in stories and symbols that present a coherent, stable interpretation of the organization. However, when an organization produces conflicting or contradictory narratives, or when the 'image in the mirror' (Dutton and Dukerich 1981) conflicts with the stories the members of an organization are telling, the organizational identity becomes destabilized (Faber 2002a).

Image and narrative in destabilized networks of higher education

This destabilization and disruption creates the discursive conditions for organizational change by creating a rationale for change. When the discourses that construct and reflect a desired identity are no longer functional, identity is disrupted. In such a network, an organization's identity must be re-formed by creating new images and narratives. These images and narratives may be constructed and situated by members of the organization – or, recalling the system model above, they may be constructed and situated by competing agents. The example of the marketization of education, introduced above, can serve again as an example of this process. The forces promoting the marketization of higher education gained legitimacy and prominence at a time when universities were struggling to define and defend their basic narratives. Throughout the 1980s and 1990s, educationalists and representatives of higher education came to realize that the university experience had been constructed in ways that alienated and disenfranchized large social groups. Research was demonstrating that large numbers of minority groups were actively being discriminated against by university entrance requirements and standardized exams, course curriculum and testing procedures, and campus community and residential activities (Giroux 1988, 1992; Rose 1990; Spivak 1993). The scholarship emerging from these issues and discussions recognized that the meta-narratives that had informed liberal education were dysfunctional. Even authors defending historical methods and content of education claimed that the narratives and images of higher education had become destabilized and dysfunctional (Bloom 1988).

During this time, when scholars and university administrators were attempting to reconstruct the narratives of higher education, economic pressures from government in the form of declining financial support for university operating budgets and declining support for research (especially in the arts and humanities) placed increased strain on colleges and universities. The image of universities reflected in the mirrors of governments, conservative think-tanks, popular media and even in the arguments of well-intentioned scholars attempting to re-legitimize higher education, portrayed places of exploitation, bias and, most crucially, of mixed and unclear educational messages and purposes. Universities became unable to successfully tell their story or present a coherent and positive image of themselves to their communities and their stakeholder groups.

Into this network of destabilized educational narratives, advocates of market-based education began creating and situating their own narratives of education. These new narratives of market-based education directly challenged the faltering narratives of traditional higher education. Hein (1999), writing about McDonald's Hamburger University, a corporate learning centre often cited as a model of market-based ('corporate') education, claimed that corporate education better responds to and integrates students' diverse cultural and economic backgrounds. Hein wrote, 'Being that McDonald's managers come from a variety of educational and cultural backgrounds, they must be taught the essentials of running a franchise' (Hein 1999, p. 75). Describing this inter-cultural context at Hamburger University, he wrote, 'half the students wear headphones as linguistic experts in glass rooms translate the instructors' words into one of five different languages' (Hein 1999, p. 75). The narrative created here tells the story of a corporation taking the education of its future multicultural managers into its own hands. As Fairclough noted (1992; see discussion above), the narrative focuses on 'skills' and 'skill training', but these meanings are shifted from traditional dualisms (skill vs. knowledge; training vs. education) to aggregate new narrative in which 'skill' represents opportunity, 'training' is advancement, and culture is a multi-ethnic corporate identity shared by all employees regardless of economic, racial or gender differences.

A second example of this competitive narrative construction can be found in a guest article written in the *Chronicle of Higher Education* (Meister 2001). Here, Meister, a consultant who helped companies create corporate learning centres, specifically challenged the relevance and necessity of traditional universities in a corporate learning environment. The article advocated for partnerships between corporate and academic universities as a way for traditional universities to protect themselves against the 'threat' posed by corporate learning sites (Meister 2001, p. 11). Meister wrote, 'the increasingly competitive higher-education environment presents a significant threat to an industry that has had a monopoly since the Middle Ages. Until very recently, colleges enjoyed a captive market, where corporations paid whatever institutions charged for executive education' (Meister 2001, p. 11). The narrative here (based on presuppositions) constructed academic universities as 'under threat', elitist, arrogant and increasingly inconsequential.[5]

The narratives created by such advocates of market-based education echo issues similar to those raised by Giroux, Rose and Spivak. In this way, the narratives of market-based education speak directly to the identity struggles experienced by traditional university educators. Advocates of market-based education have structured their narratives within the space contested by higher education to argue that the corporate university is more egalitarian, merit-based, efficient and practical than traditional higher education.

These contesting narratives demonstrate that such stories are not about truth or accuracy. On a functional and semantic level, they are stories that compete for situational meaning and legitimacy. At the same time, structurally and pragmatically, these narratives can also be interpreted as part of what Berkenkotter and Huckin (1995) called 'genre knowledge' because the texts use conventions (narratives, images, vocabulary) from within the targeted social group to attempt to make changes (disrupt, destabilize, restabilize) within that same group. Genre knowledge is central to understanding the process of change because, as Graham argued, the 'level of genre is where institutional ructions are first expressed' (Graham 2004, p. 62). It is at this level that that text, discourse and social mediation intersect and contest extant knowledge and practice.

Genre knowledge: enabling transitions from the familiar to the new

Berkenkotter and Huckin wrote that 'genres are inherently dynamic rhetorical structures that can be manipulated according to the conditions of use' and that 'for writers to make things happen . . . they must know how to strategically utilize their understanding of genre' (Berkenkotter and Huckin 1995, p. 3). In conditions of change, writers must be able to strategically utilize the predominant genres of the social network in an attempt to change that same network. Thus, if narrative and image provide the means for disrupting social networks and creating dysfunctional identities within an organization or social group (the subversive displacement of meaning), genres can provide key mechanisms for the (re)construction and stabilization of new representations.

Genres enable transitional change because they are dynamic, situational and strategically value-oriented. Graham noted that genres elicit and solicit social expectations (Graham 2004, p. 56), and 'because they are patterned ways of producing expectations, genres link social pasts with the present, and with possible futures' (Graham 2004, p. 54). Following

Bourdieu and Wacquant's concept of 'habitus' or regularized, habitual activity (1992, p. 139), Schryer noted that genres are 'constellations of regulated, improvisational strategies triggered by the interaction between individual socialization ... and an organization' (2000, p. 450). Genres are regulated because they emerge from a consensus of everyday practice and are recognized and trusted as a relatively formal part of that structure. Yet, as Schryer specified, genres are still performed improvisationally and it is this characteristic that enables agents to use genres for transitional purposes.

The functional tension between regulation and improvisation makes genres powerful written instruments of change. As familiar representations, genres inspire trust and are linked to the symbolic capital of social groups (Bourdieu 1977, p. 179). As forms of symbolic capital, or what Bourdieu also called 'credit', the genre is part of a system of belief and trust within social groups (Bourdieu 1991, p. 192). This trust can be leveraged to create acceptance during the social change transition. Genres provide writers sufficient stability and trust to link to the old while simultaneously restructuring (re-aggregating) the new.

Berkenkotter and Huckin described the product of this re-aggregation as a 'generic blend' (Berkenkotter and Huckin 1995, p. 114), when writers combine features from two or many subgenres to create a new genre or stabilize an existing one. Alternatively, a new aggregation could be identified as a 'hybrid' genre (Latour 1999; Spinuzzi 2003, pp. 160–1; Winsor 2003, pp. 140–1), a combination of different structures and events that produce a distinctly new approach or event while still resembling the 'parent genres' that produced them (Spinuzzi 2003, pp. 160). Genre hybridity is especially relevant from an activity perspective (Graham 2004, p. 64) since, as Fairclough argued, 'social change is change in the networking of social practices' (Fairclough 2004, p. 112). Those practices that constitute (new) genres are often formed in networks of necessary activities because of the limitations of current forms of action (Spinuzzi 2003; Graham 2004). Fairclough used the term 'order of discourse' to describe 'a particular articulation or configuration of genres, discourses, and styles' (Fairclough 2004, p. 112). Further noting the reciprocal relationship between language and social network, he wrote that orders of discourse are the 'social structuring of semiotic difference or variation' which is mediated by interdiscursive links among various texts. Thus, a particular mix of genres, discourses and styles is evident in specific text features but also in the way the text works as a social practice (Fairclough 2004, pp. 112–13). In this way, social practice may elevate specific texts, diminish others, and leave others vulnerable to collocation and hybridity.

Genre knowledge and generic hybridity in corporate university networks

This process of constructing generic hybridity from orders of discourse was described by Connell and Galasiñski (1998), who traced the creation of academic mission statements in UK universities (higher education institutions, HEIs). They noted that the mission statement is a socially powerful genre that has emerged from market-based ideology and they anticipated that some may see the imposition of mission statements within higher education 'as another symptom of the "marketization" of post-compulsory education, of its incorporation into the commodity market ... and the general reconstruction of social life on a market basis' (Connell and Galasiński 1998, p. 457). However, Connell and Galasiński showed that the writers they studied created hybrid documents that used the genre of the mission statement but advocated for the independence and authority of the university apart from the market. They wrote that the statements:

> did not always, nor did they only, set out to represent academic affairs as 'business like'. There is evidence that the Statements both acknowledged *and* negotiated the political-ideological network. In some cases, at least, while they made lexical concessions to what may be regarded as commercial discourses, they still represented core academic affairs much as they have always represented them.
>
> (Connell and Galasiński 1998, p. 476)

The hybrid discourse combined discourses of the market with discourses of traditional academic practice to create a negotiated stance that Connell and Galasiński wrote was based primarily on strategic ambiguity and an absence of precision. They argued that 'excellence is about the only state which the HEIs are prepared to say they have achieved and will maintain ... Yet the (deliberate) uncertainty about what will be delivered and when allows them to keep their options open, for who knows what their paymasters will expect of them in future' (Connell and Galasiński 1998, pp. 476–7).

The *strategic ambiguity* described by Connell and Galasiński has been found to be a generic characteristic of the new capitalism as constructed by advocates of market-based practices (Swales and Rogers 1995) and advocates of market-based higher education programmes (Faber 2003). For example, Meister's argument (above) is constructed through modality, presuppositions, hedges and amplifications. She writes, 'a number of corporate universities are offering courses not only to their direct constituencies, but also to the public' (Meister 2001, p. 10). This claim rests on the ambiguous 'a number', the temporally ambiguous 'are offering', and the ambiguous nouns 'courses' and 'public'. Similarly when Hein (above) transitioned from simply describing Hamburger University to generalizing the benefits of corporate-based education, he also relied on hedges, predictive verbs and presuppositions. For example, he wrote, 'Other companies, especially during the course of the last decade, have realized the value of corporate universities as well' (Hein 1999, p. 76), using the hedge 'other', and the presupposed value of corporate universities. Similarly, the sentence 'the creation of corporate universities has become a major trend over the course of the last 10 years' (Hein 1999,

p. 75) also uses nominalization (the creation), hedging (major) and temporal ambiguity (has become, over the course of the last 10 years) to construct sufficient strategic ambiguity to insinuate a market-based discourse as a destabilizing agent within vulnerable academic contexts.

Theoretical backgrounds: social change as discursive practice

Before moving to explain more micro-level features of change discourse, I would like to briefly pause this introductory chapter. Here I will step back to explain the theoretical orientation that informs this project and propose a rationale for examining social change at more micro, syntactic and pragmatic levels.

This book emerges from the critical study of discourse, which has as its principal goal the study of language in practice. 'Critical linguistics' emerged in the mid 1970s as a method for studying functional language, meaning language as it occurs in day-to-day operations. What differentiated critical linguistics from other forms of language study was a desire by researchers to understand the connections between language and ideologically driven social formations (Fowler 1996, pp. 3–4). In addition to learning about the deployment of specialized forms of language in social situations like political discourse, media coverage and professional advocacy in medicine or law, these researchers were interested in examining and explaining how language works in day-to-day networks to achieve specific social actions. This is what Jay Lemke has called 'the social activity of making meanings with language and other symbolic systems in some particular kind of situation or setting' (Lemke 1995, p. 6). For researchers working within this area, language is not separate from its social function. To use Kress's term, language is not a 'client' of other disciplines (Kress 1996, p. 15), a simple way to evoke content. Language is an essential, integrated and ethical function of meaning-making. 'Discourse' is used to describe language as it is used to achieve specific purposes.

What has intrigued scholars in the field of critical linguistics and critical discourse analysis has been the property of discourse to reflect and construct meaning simultaneously. An effective discourse fits within a particular situation; yet, it simultaneously is used to construct that situation and successfully vie with other potential discourses to give individuals and groups meaning, legitimacy and power (Gee 1999, p. 11). This constituted/constituting dynamic of modern discourses warrants an analysis that connects discourse in its performative role to its strategic function within the operations of knowledge politics.

My project also emerges from the disciplinary context of rhetorical studies which claims a long history of oratorical and textual study reaching back to the Greek Sophists and Aristotle's definition, 'let rhetoric be [defined as] an ability in each [particular] case, to see the available means of persuasion' (Kennedy 1991, p. 36). Modern rhetorical studies of organizational writing have perpetuated and extended this tradition by emphasizing ethnographic, participatory and qualitative research of the contexts of document production and the reception of specific documents by users and other audiences (Giltrow 1998; Schryer 2000). And, over the past two decades, rhetorical studies of workplaces have produced sophisticated understandings of the context of document production and the ways social forces influence text production and reception. However, where the rhetorical tradition has been less successful has been in the examination of specific discourse features and a precise accounting of how micro features (syntax, pragmatics) enable and realize macro features (genre, ideology, social action, ethics) (see Giltrow 1998). As Schryer has argued, 'what we lack is an overall framework', which combines the micro analysis of linguistics with the macro study of rhetoric (Schryer 2000, p. 461).

There is sufficient history in this current stalemate to warrant a brief tangent. The theoretical framework established by social constructivism from the mid 1960s to the 1980s provided a rising tide for communication scholarship, captivating projects in English studies, rhetoric, sociolinguistics, discourse studies and communication. The now taken-for-granted idea that knowledge is formed by social communities enabled researchers to claim that language, and more precisely writing, was a major force in the social construction of this knowledge. As Bazerman and Paradis wrote, 'writing is more than socially embedded: it is socially constructive. Writing structures our relations with others and organizes our perceptions of the world' (Bazerman and Paradis 1991, p. 3). From this perspective, several generations of scholars claimed that by examining a text within its social network we examine and understand the means by which reality is constructed.

This work has largely fulfilled its purpose, which was to demonstrate that texts are part of the social construction of knowledge. But, a problem has emerged in that we cannot take this functional claim – that texts are part of the social construction of knowledge – and derive from it a more substantial claim – that we understand how texts construct social knowledge – or an an ethical claim – that we know how texts fulfil or promote values-based activities. In other words, what is limiting a fuller understanding of constitutive discourse is a way to examine such discourse as it occurs, what Kelly-Holmes called the 'how' of what texts are saying (Kelly-Holmes 1998, p. 343). Bazerman and Paradis claimed that 'by understanding texts within the professions, we understand how the professions constitute themselves and carry out their work through texts' (Bazerman and Paradis 1991, p. 1). But, simply claiming that the professions (or workplaces) constitute themselves textually does not adequately fulfil this goal. We know that workplaces, organizations and the professions constitute themselves and carry out their work through texts. But at the micro level of particular discursive phenomena, we do not know how this happens. And, this is a crucial distinction.

Schryer, in her work on negative messages in insurance letters, proposed that genre study can be a locus for integrating rhetorical study with critical linguistics. She noted that, historically, genre researchers have worked in either a rhetorical or a linguistic tradition, with rhetoricians emphasizing the impact of texts and linguists the details of specific documents (Schryer 2000, p. 451). Working from Bourdieu and Wacquant's concept of 'social praxeology' (Bourdieu and Wacquant 1992, p. 11) Schryer proposed, first, reading texts 'that instantiate genres to describe and critique the strategies that some genres activate to represent power' (Schryer 2000, p. 461) and, second, 'acknowledging the practical logic or problem-solving techniques' agents will employ that may not fully comply with established generic patterns (Schryer 2000, p. 461). Together, these two steps enable an examination of standardized features and improvised acts, what Bourdieu and Wacquant called 'feel for the game' (Bourdieu and Wacquant 1992, p. 128).

Like Schryer's non-compliant agents, social change also presents discursive moments that tend to pre-figure or resist stabilized discursive forms like genres. Although genre research has broadened formative concepts to include less formalized discursive forms such as social action (see, e.g., Berkenkotter and Huckin 1995) or temporary alignments (Schryer 1993), contexts of social change are typically conscious attempts to resist, destabilize, and eventually replace existing genres. Change seems to become a possibility when existing genres are violated (Faber 2002a, p. 172). For these reasons, my own project does not focus so much on genre or the descriptions of specific genres as it seeks to understand genre activity, hybridity and dissolution in the face of social change. Nor have I found it necessary to question whether discourse enables or structures social change. Following Bazerman and Paradis, I take the discursive construction of social networks to be a basic assumption. Instead, my purpose is to focus on and articulate the function of discourse within the nexus of change as new social systems and relations solidify to structure a newly emergent social space. To borrow a phrase from Halliday and Martin, I have constructed this project to 'reason grammatically' about the process of change (Halliday and Martin 1993, p. 22) and thereby examine and theorize the intersections of ethics and discourse in modern contexts of change.

As inferred above, I agree with Chouliaraki and Fairclough that social action is mediated by discourse. As an activity (such as change) is articulated either by speech, writing or even gesture, the links between discourses and social positions become more fully established and evident (Chouliaraki and Fairclough 1999, p. 150). To fully examine how ethics and values are being re-articulated in modern discourse, alongside Kress and Hodge's work (1993) I have employed a version of Halliday's Systemic Functional Linguistics (SFL). SFL positions language as a resource for making meaning rather than as a system of rules. It is a descriptive rather than a purposefully normative orientation concerned with identifying the relations between texts and their social contexts (Halliday and Martin

1993, p. 22). This orientation encourages researchers to see texts as having multiple levels and multiple roles in any given social network. For example, a text may influence public understanding of government policy while simultaneously serving as a marketing device for an electoral campaign and a refutation of a prior administration's policy. As texts enact these roles, they actively construct meaning rather than passively act as a conduit for thoughts or feelings. Although thoughts and feelings will influence the production of texts, interpretation and the construction of meaning occur independent of these originary influences as they occur in different contexts and different times.

Functionally, researchers using an SFL model work to identify the various roles constituent parts play within a selected text. Rather than document a noun as solely a part of speech, for example, an analysis may indicate the thematic function played by the noun and the noun's role in introducing new, as opposed to given, information. Such an analysis will operate by documenting the repetition of a constituent function (such as a noun, verb, adjective) at grammatical and pragmatic (functional), semantic (content) and contextual levels (discussed below as textual, interpersonal and ideational functions, Halliday and Matthiessen 2004). For example, a noun's occurrence within the thematic (left-most) position within a phrase is interpreted by that same noun's cohesive function as part of a larger text, the semantic meanings (definition) it makes available, and the contextual references to which it points. These discursive levels are offered as three integrated functions: the textual function, the interpersonal function and the ideational function (Halliday 1994; Bloor and Bloor 1995, pp. 7–9; Chouliaraki and Fairclough 1999, p. 140; Lemke 1995, p. 41; Halliday and Matthiessen 2004, pp. 29–30). In brief:

- *The textual function* articulates the grammatical or organizational function of language. The textual function articulates 'the construction of relations between elements of the discourse itself' (Lemke 1995, p. 41). This function interprets the structure, texture, cohesion and the forms by which the information is organized within specific clauses in the text. Textual information is obtained through an analysis of the grammatical (syntax) constituents of the clause and then correlated with interpersonal and ideational functions.
- *The interpersonal function* articulates how the text orients and positions its audience, its originator and the relationships between these actors. This function examines the social relations and social identities enacted within and across the constructions generated by the text and its references (Chouliaraki and Fairclough 1999, pp. 60–1).
- *The ideational function* articulates the propositional content of the text (Huckin 1992, p. 87). Through its ideational function, the text articulates an interpretation of the world, what Chouliaraki and Fairclough call the 'practices' of social life (Chouliaraki and Fairclough 1999,

p. 21), and it constructs 'how things are' through its explicit descriptions and intertexual references (Lemke 1995).

In a broader approach, one could situate these as an examination of the various practices the discourse enacts (Chouliaraki and Fairclough 1999, p. 61), especially when these practices entail aspects of social control, influence or motivation. The textual, interpersonal and ideational functions bridge micro and macro aspects of discourse as they link micro (syntax, pragmatics) constituents of the text to contextual influences and determiners. Chouliaraki and Fairclough called this identifying the 'ideological effects of discourse' (Chouliaraki and Fairclough 1999, p. 150), noting that 'the social is built into the grammatical tissue of language' (Chouliaraki and Fairclough 1999, p. 140). In other words, the three functions operate intertextually with other grammars and other texts. Even in their grammatical structure (system) they may integrate material activities and constraints, existing social relations, or mental models (Lemke 1995, *passim*; see also Chouliaraki and Fairclough 1999, p. 61).

Given that 'all meaning is intertextual' (Lemke 1995, p. 41), words, phrases and clauses have meaning potential in relation to the other phenomena in which they occur and to which they refer. In addition, the pragmatic and grammatical features of phrases and clauses also create meaning potential in the ways they position and realize various other internal and external components of the text. More recently, Fairclough (2003) augmented this three-fold system of meaning construction by enfolding Halliday's terms within what Fairclough called 'types of text meaning': genres/actions, discourses/representations and styles/beings. For Fairclough, genre distinguishes ways of acting, discourses ways of representing and styles ways of being (Fairclough 2003, p. 26). Though the terms 'genre', 'discourse' and 'style' can be points of dispute in their respective disciplines, using these terms to identify ways of acting, representing and being (identity) through discourse provides a way to characterize discursive function as language simultaneously represents and constructs meaning, builds and enacts social relations among various actors within and between specific social networks, and provides an internal logic and coherence in its own textualization (see Fairclough 2003, p. 27; Chouliaraki and Fairclough 1999, pp. 63, 151).

Figure 1.1 outlines Fairclough's model and Figure 1.2 depicts a conflation of Halliday and Fairclough's models using the terms action (genre), representation (discourse) and identification (style).

Readers familiar with Halliday's model will recognize connections between his terms and Fairclough's articulation of action, identification and representational types of meaning. As Figure 1.2 depicts, Fairclough has combined Halliday's textual and interpersonal functions to characterize action, accentuating the formal function of texts in constructing and representing activity (Fairclough 2003, p. 27). At the same time, the

Fairclough's types of text meanings
Action: interpersonal and textual: genres
grammar, texture, structure, cohesion

Representation: ideational: discourses
interpretation of the world, propositional content

Identification: social and textual function: style
social relationships enacted by the text

Figure 1.1 Fairclough's types of text meanings (adapted from Fairclough 2003, p. 27)

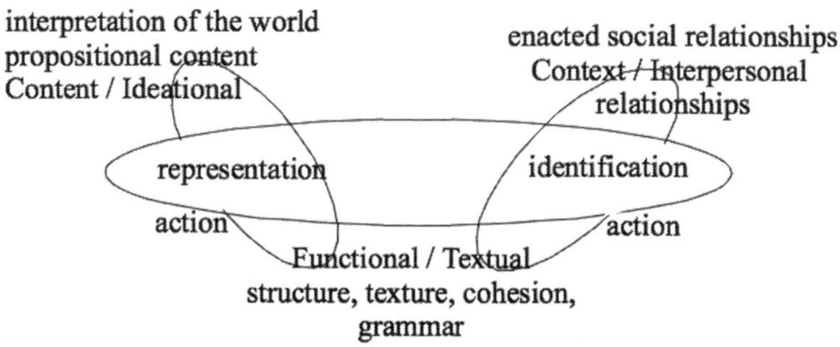

Figure 1.2 Conflation of discourse models

textual metafunction should not be excluded from what Fairclough called 'identification' because textual features influencing cohesion, comprehension and what is included and excluded from texts are key dynamics of personal and institutional styles. For this reason I have placed identification in a reciprocal relation with action constituted by textual and interpersonal meanings. At the same time, extending Fairclough's model, I have also placed action in constituting relations with representation since representational meaning constitutes a different form of action.

Perceiving meaning construction as building relationships of action-representation and action-identity is useful in describing a fairly stable social network. Yet, when trying to examine or predict discourse activity in activities enacted to promote social change, action, representation and identity are mediated by immediate and absent-but-influential discourses. Linguists use the term 'syntagmatic' to point to relationships between and among items present within a text (Fairclough 2003, p. 37; Bloor and Bloor 2003, pp. 241–2; Hodge and Kress 1993). Syntagmatic analysis enables us to examine relationships between agents within a discourse and the ways these agents are represented and influence other agents and objects around them. 'Paradigmatic' relations point to relationships between and among words that are present and words that are not. For example, a reporter may choose to use the word 'discuss' rather than 'fight' to describe an event between two people. Or the reporter may decide to elide that specific encounter and instead focus on another aspect of the event. Paradigmatic relations express choices and values and can be examined for important value- and ideological-based decisions by the author. For example, Huckin characterized specific ways writers omit information in order to create biased presentations of social events. Huckin noted that such 'silences' can be manipulative when accomplished to make an ideological argument (Huckin 2002, 2004).

Hodge and Kress (1993) presented what they called a 'transformational model' to examine relationships between the syntagmatic and paradigmatic discourses and transformations such constituents enable within and between texts. For example, nominalized verbs (e.g. 'the critics' *reception* of the play' [Quirk *et al.* 1985, p. 1289]), passive constructions, questions, conversions, negative constructions and other diversions from typical cohesion patterns signal paradigmatic choices that influence meaning construction. These paradigmatic choices enable social transformation and change. Paradigmatic choices are initiated with persuasive purpose and theoretically, an identifiable ethical stance. For example, Hodge and Kress demonstrated that as language transformations, certain sentence level (functional/action) text features are associated with economy, efficiency (simplifying process by excluding detail) and distortion (intentionally excluding important information) (Hodge and Kress 1993, p. 10). Interpersonally (identification) these features may signal a paternalistic and condescending relationship between the speaker and audience. Thus, even before specific content (content/representation) is even considered, grammatical and pragmatic features have signalled value-oriented actions initiated by the speaker and the audience.

Discourse (micro)-based studies of the text-of-change

The more micro, grammatical level of examination detailed above represents a relatively recent interest among researchers examining the syntactic and pragmatic features of specific texts of change. Such examinations of the textual features of change have suggested key grammatical and pragmatic features of change-enacting texts. Specifically researchers have examined grammatical metaphor, theme/rheme relationships, modality and presupposition as textual features commonly associated with social change. For those less familiar with this terminology, Table 1.1 provides a brief introductory definition and example of the linguistic terms used in the following discussion.

Such work has emerged from a relatively new effort within writing research to combine Systemic Functional Linguistics (SFL) (Halliday 1994) with Critical Discourse Analysis (CDA) (Fairclough 1992, 1995; van Dijk 1995, 1997 (ed.), 1998; Wood and Kroger 2000; on combining

Table 1.1 Definition of linguistic terms used in Chapter 1 discussion

<table>
<tr><th>Term</th><th>Definition</th><th>Examples</th></tr>
<tr><td>Action processes</td><td>accent accomplishments and realize processes</td><td>We create a new system
She completes the exam
We manage change</td></tr>
<tr><td colspan="3">Action processes are significant because they constitute the 'noun' as real and situate the verb phrase as a concrete process (rather than a possibility). For example, the phrase 'manage change' implies that change is real (has already happened) and places it within an ongoing process – management. The action process could be seen as a device for presupposition since negating the phrase does not negate the existence of the noun: 'do not manage change' → the negation does not influence the existence of change.</td></tr>
<tr><td>Presupposition</td><td>grammatical construction that requires audience to admit existence (reality) of one constituent within the clause.</td><td>Have you stopped cheating?</td></tr>
<tr><td></td><td>Presuppositions are evidenced by negating the claim. The presupposed entity survives the negation.</td><td>We have not found the water in the desert (there is water in the desert)</td></tr>
<tr><td>Passivization</td><td>use of a passive verb</td><td>A new system was created
The exam was completed
Change was managed</td></tr>
<tr><td colspan="3">Passivizations are important because they remove agency from the phrase to create the appearance of actions that occur without or beyond human intervention.</td></tr>
<tr><td>Nominalization</td><td>turning a verb into a noun</td><td>The creation of a new system
The completion of an exam
The management of change</td></tr>
<tr><td colspan="3">Like passivization, nominalization removes the social actor from the assertion.</td></tr>
<tr><td>Grammatical metaphor</td><td>systemic term for nominalization, usually situated as a proposition. Requires a grammatical substitution in which one grammatical function (e.g. noun) is substituted by another (e.g. verb)</td><td>Creating a new system
Completing an exam
Managing change</td></tr>
<tr><td colspan="3">See nominalization (above).</td></tr>
</table>

Term	Definition	Examples
Theme/rheme	Usually, **theme** designates the left-most constituent of a clause. *Rheme* designates the constituents after the verb.	**We** created *a new system* **She** completed *an exam* **We** manage *change*
Given/new	**Given** information is usually associated with the **theme**. Clauses usually present given information in the initial position. *New* information is usually associated with the *rheme* and introduced in subsequent positions.	**We** created *a new system* (assumes recognition of 'we' – 'a new system' assumed new)
Genre hybridity	combination of two or more genres into a new genre. Usually considered a hybrid while the new genre is still temporary.	Web based instructional information + advertising. = web-based info/ad
Strategic ambiguity	discourse features that create ambiguity of meaning but also provide a 'key' to resolution which privileges an ideological position (typically created by modality, progressive aspect)	We *are creating* a new system (when started? finished?) We *can* manage change (under what conditions?)

Systemic Functional Linguistics with Critical Discourse Analysis see Fowler *et al.* 1979; Fairclough 1989; Martin 2000; Young and Harrison 2004). Chouliaraki and Fairclough wrote that Critical Discourse Analysis:

> would gain from extending its so far limited relationship with SFL (essentially just using a systemic grammar of English as a method in text analysis), not only in terms of using SFL as a resource for analysis but also towards a theoretical dialogue over such issues as the relationship of semiotic to social change, and the nature of what systemicists have called 'semologic' (Hasan [1999]) – the generative power of the semiotic, or its generative 'mechanism' in the terms of critical realism.
>
> (Chouliaraki and Fairclough 1999, p. 139)

Combining Critical Discourse Analysis with Systemic Functional Linguistics provides a robust analytical framework for examining the interconnectedness of language and social network and specifically the interactions of language, social network, and social change. Such studies have pointed to specific semiotic, syntactic, and pragmatic features found in texts of change and may eventually help to detail specific discursive conditions that enable social change. Hasan (2004) noted that a major task for this work will be to determine the relationship between discursive variation and social change. Understanding variation will enable researchers to place specific activities of change within specific types of discursive variation. However, as Hasan claimed, research has not yet explained this relationship. She wrote that there is 'next to no reliable account of how the "genres of power" grow' and no real understanding of how specific forms of language use have given rise to specific states of affairs (Hasan 2004, p. 43). To better attain this knowledge, researchers must examine 'systemic stability' and how it is maintained. She wrote, 'to understand stability is to know what hinders change; to understand system is to understand why process so overwhelmingly tends the way it does' (Hasan 2004, p. 45).

Further, Bartlett cautioned that a linguistic description of change can not be formed from a few examples of variables in use. When coding linguistic variables that appear to be associated with discourses of change, researchers must be careful to locate large numbers of uses and from these uses identify common meanings and functions that unite such use (Bartlett 2004, p. 71). This caution is important because, in moments of change, texts and social contexts are always in reciprocal relations. Fairclough has reminded, in moments of change, 'the "overdetermination" of language by other social elements becomes massive' (Fairclough 2004, p. 116). This makes it extremely difficult and problematic to state with certainty that specific linguistic features universally determine certain social actions or that specific social actions will always initiate specific textual features. Text and context are mutually informing and dynamic.

Working within these limitations and cautions, several researchers have attempted to articulate and empirically demonstrate specific discursive variables associated with changing social networks. These studies either assume or demonstrate the theoretical perspective outlined above – that change occurs as the resolution of strategic ambiguity. Issues and problems informing this work focus on how such ambiguity is discursively constructed within a specific network, how such texts and their writers achieve and maintain systemic stability, and how to empirically identify the discursive features that enable or support this process.

Harrison and Young (2004) reported on major announcements made by a senior manager in a Government of Canada department (Health Canada) regarding changes he was instituting to the department. The site for this study was a new branch created by the senior manager and the first piece of official communication from its new head to the 1,200 staff amalgamated from several other Health Canada units. The communication (memo) was the manager's attempt to stabilize the changes he had made within the department and introduce the future direction of the new branch. Harrison and Young's study showed that the memo emerged as a hybrid of different managerial styles. The manager attempted to convey a more egalitarian managerial style within the new department with discourse marked by references to sharing power, delegation of decisions and local control. However, integrated throughout this discourse were forms of discourse representing 'command and control' management (Harrison and Young 2004, p. 235). Harrison and Young showed that while much of the semantic content of the memo suggested the more egalitarian style, grammatical and discursive features enacted a command and control structure that resisted the semantic content and became the most prominent feature of the memo. In this case, the command and control discourse was enabled by strategic ambiguities that negated the egalitarian propositions.

Table 1.2 presents features that created this strategic ambiguity as described by Harrison and Young.

Table 1.2 Discursive features creating strategic ambiguity

Verb choice: Action processes (e.g. 'work', 'make', 'set', 'present') which enhanced accomplishments and made them seem to be established and finished activities (Harrison and Young 2004, p. 237).
Passivizations combined with nominalizations which discussed background decisions that had not yet been finalized and which removed agency from these pending decisions (Harrison and Young 2004, pp. 237–8).
Grammatical metaphor: Nominalized processes in the subject position that obscured agency and made activities appear completed.
Nominalizations that concealed the fact that decisions were made without staff input (Harrison and Young 2004, pp. 238–9).
Theme construction with given information: Themes announced decisions demonstrating that 'new' information was assumed to be taken for granted (Harrison and Young 2004, p. 239).
Genre hybridity: Traces of advertising and promotional genres within bureaucratic discourse (Harrison and Young 2004, p. 239).

(Aggregated from: Harrison and Young 2004)

Harrison and Young concluded that these pragmatic features obscured the fact that some aspects of the change remained unknown. Secondly, since much of the discourse created ambiguity about the process and future direction of change, the memo placed the senior management clearly in command and control positions directing the creation and the implementation of the new branch. Harrison and Young argued that the memo was a rhetorical failure because it alienated many staff, 'creating staff resentment and resistance to the hierarchical status quo reinforced by the discourse' (Harrison and Young 2004, p. 242). This failure does not appear to be due to the creation of ambiguity, but because the manager was unable to successfully resolve this ambiguity in a way that was not based on his own power.

The creation of strategic ambiguity presented by Harrison and Young was also prominent in Harvey's (2004) study of transformational leadership discourse at Apple Corporation. Examining speeches by Steve Jobs, CEO of Apple Computer, Harvey pointed to specific text features that enabled Jobs to construct a potent and compelling forum of contextual ambiguity which allowed him to instigate and compel his own social change within Apple. In some ways answering the conclusions resulting from Harrison and Young's study, Harvey wrote that the 'ability of transformational leaders to motivate their employees to "take on board" and pursue a particular vision depends to a large extent on the leader's rhetorical skills' (Harvey 2004, p. 247).

In her description of Steve Jobs' transformational discourse, Harvey showed that initially his speeches proceeded by creating ambiguity but then resolved that ambiguity by appealing to what she called his 'followers' self-concept' (Harvey 2004, p. 248). Jobs frequently used verb constructions that relied on do + ____ as in 'do this', 'do something' or 'do nothing'. The significance of this construction lies in the abstraction of 'do' as a verb. Jobs emboldens his employees to 'do', to take action, but the specifics of that action are often vague or abstracted by metaphor ('we have to *drive a stake in the ground*'; 'all the work we've done we throw down the toilet' [Harvey 2004, p. 252]). These metaphors enabled Jobs to cite the symbolic prospect of action without necessitating particular activities or proposals. In addition, these metaphors characterized change as inevitable and irresistible, instantiating 'windows of opportunity', 'certain realities' and required actions (Harvey 2004, p. 252). Finally, Jobs used collective pronouns (we) which de-emphasized his own agency (Harvey 2004, p. 253) and made the accountability for change appear communal and organization-wide.

Like the features discussed by Harrison and Young above, abstract action references ('do'), metaphors and collective pronouns also appear to be strategies that can discursively destabilize an existing network. Harvey extended her analysis though to suggest features that helped Jobs to restabilize his network. Harvey argued that this restabilization was attempted through overt appeals to employees' identity, efficacy and sense of self-worth (Harvey 2004, p. 254). Jobs argued that in the face of uncertainty, his employees have shown a great capacity to work, tenacity in the face of struggle, and an ability to embrace opportunities (especially those that fate provides) (Harvey 2004, pp. 255–6). These characteristics (values) identify a self-esteem that Jobs constructs and recognizes within his employees. This identity-construction enabled Jobs to create two distinct camps for his followers, those who would succeed (overcome) and those who would fail. Harvey saw in this division the movement from collective unit to individual decision-maker, from group to specific agent. She wrote, 'It is here that the organizational identity breaks down into individuals; Jobs is not prepared to tolerate faintheartedness, i.e. insecurity, and the talk of failure is not an option' (Harvey 2004, p. 257). After constructing situational ambiguity, Jobs forced followers to make a decision to embrace his leadership and his changes or give in to failure.

Finally, in a study of the promotion of genetically modified foods, Lassen (2004) suggested a linguistic-based research focus could help researchers better understand and document the 'coherent link' that has been theorized as a transitional bridge to the new. Lassen (2004) hypothesized given/new relationships in texts of change, suggesting that the interplay between theme/rheme and given/new may form a transitional textual path for social change. Halliday (1994) characterized theme as the information in a clause that precedes the verb and rheme as the information that pertains to the subject of the clause but comes after the verb. Typically, themes will present given information (information that is assumed to be known or familiar to the readers) while information that is assumed to be new for the reader is presented in the rheme.

Examining texts about genetically modified food, Lassen showed that clauses 'developed their themes by frequently resuming an earlier rheme as a theme in a new sentence' (Lassen 2004, p. 270). Writers introduced new material in the rheme and then, once introduced, this new material was then used to construct new themes in the text. Lassen argued that treating new information in one sentence as given information in a subsequent sentence 'has the effect that its information can no longer be challenged' (Lassen 2004, p. 270). Further, Lassen noted that ideational grammatical metaphors were used to resume the rheme as a theme in the subsequent sentences. In other words, rhemes were nominalized and their agency was eliminated to make the assertion appear completed in the thematic position. For example, the rheme 'are part of the company's ongoing commitments to global agricultural research' was turned into the theme 'Monsanto's commitment' in a later sentence. This strategy enabled the author of change to achieve cohesion across a transitional text and create a coherent link between old and new ideologies.

Engaging and critically examining the discursive moves that enable social change requires a multilayered accounting of both discourse practices and the ideological values reflected and constructed within such practices. As such, it requires a unified study of micro and macro constituents of discourse. To accomplish such a task, this project proposes taking an overt textual focus to the moment(s) of change in modern organizations.

Discursive stability and social change: dynamic and synoptic style

Transformation also denotes movement from paradigmatic to syntagmatic instantiation as speakers/writers make absent discourses present. This process is formulated in the next chapter as the creation of discursive stability or what Bazerman called 'stabilized representations' (Bazerman 1999, p. 308). Syntagmatic and paradigmatic activities point to relationships and patterns of presence and emergence as specific discourses are transgressed, innervated and displaced by other competing (and cooperating) discourses. To claim that a certain discourse is 'stable' is not to suggest that its originator is fabricating truth or is acting disingenuously. 'Relative stability' describes the confidence writers or speakers appear to have in the claims being made. This is not so much a lie detector test as it is an evaluation of how well ensconced a claim appears within an unfolding discourse. As always, there are some assumptions here: that the speaker or writer is not intentionally trying to deceive, that claims are offered in good faith, and that, to a degree, statements are not strategically manufactured (spun). At the same time, even with dishonest and manufactured discourse, coherent meaning-making is restricted by textual, interpersonal and ideational context which limits the extent to which unrelated and fabricating information can be included within a discourse.

Working from studies of scientific discourse, Halliday and Martin (1993) identified specific characteristics of texts as more or less stabilized within their enfolding network. Halliday and Martin used the term 'dynamic style' to describe highly-claused texts with cohesive structures that framed or situated meaning as current, ongoing and under way (Halliday and Martin 1993). These structures represented events as fluid, in-process happenings. The 'dynamic style', as a form of cohesion, suggests that events portrayed within strings of clauses are relatively unstable since they required several degrees of modification.

Whereas dynamic texts unfold in elaborate clausal structures, Halliday and Martin argued that texts with relatively few clauses exhibited a degree of abstraction from fluid, ongoing accounts. They argued that these texts represented generalizable and completed claims and events. Halliday and Martin called these structures the 'synoptic style', claiming that they reflect completed events, firm categories and static structures (Halliday and Martin 1993, pp. 46–9).

Situating claims about the relative stability of transformative discourses within the dynamic/synoptic framework developed by Halliday and Martin, we can posit that the dynamic style, indicating an activity in-process, represents texts/events as less stable than a synoptic style that reports on a completed, abstracted event. Dynamic texts would be more closely aligned with destabilized networks (social networks undergoing change) while synoptic texts would be more closely aligned with stabilized networks (or networks authors would prefer audiences to interpret as stabilized).

By making this juxtaposition, I am not commenting on an author's commitment to the intended results of a specific claim nor am I addressing the author's (or others') perceptions and beliefs about the relative truth of the claim. Instead, dynamic and synoptic styles instantiate the degree to which a textual network (a changing social system) can be said to be completed, stable and thereby able to be abstracted rather than fluid, changing and thereby still 'in play'. Networks of change require the progressive stabilization of discourses which can be characterized as the movement from the dynamic (fluid and modified) to the synoptic (completed).

This rather admittedly complex theory of meaning production is important because it characterizes discourse as a social activity. Examining discourse production as a social activity necessarily denaturalizes the appearance of texts as autonomous entities and thereby re-aggregates language with its ethical composition. This new aggregation of discourse and ethics, texts within contexts, does not reconstitute the individual as the sole architect of discourse. Nor does it rest final responsibility with the individual. Instead, it views individual action as a production and a consequence of social discourses. Thus, individuals are equally constituted by discourse and subject to the dual constituting/constituted role of language. At the same time, people can be held accountable for recognizing and understanding their own discourses and the ways these discourses operate within social networks.

This book turns as its focus to the transformations (and other features) of language in networks of change because change represents dynamic, unstable and volatile moments within organizations (Faber 2002a, 2003). Change entails contests for meaning, legitimacy and power and it is during these contexts of transformation that ideological issues come to the fore (Connell and Galasiñski 1998; Kelly-Holmes 1998), thus making for a rich environment to connect the research questions of both rhetoricians and linguists.

Bazerman wrote, 'information is neither disembodied nor neutral. Information is rhetorical. It is created and deployed in particular historical circumstances for the use of particular individuals and groups' (Bazerman 2001, p. 290). Change is caused, enacted and enabled by both material and language events. Discourse is one part of the dialectic that informs – and is informed by – social action.

Acknowledging the dialectic between material conditions and discursive action (that material conditions influence discourse practices and that discourse influences material conditions) does not reduce or disparage the importance of discourse to social life within organizations. Similarly, acknowledging that human action is influenced and constructed by discursive formations does not remove us from ethical accountability. Instead, it asserts that as an equal component of social life, discourse necessitates an equally rigorous interrogation as economic, physical or other material aspects that inform organizational decision-making and the social life of organizations. This position asserts that discourse-based research in rhetoric and critical linguistics needs to better examine and understand the connections and influences between and among discursive and material phenomena. In an effort to contribute to this task, this book offers a renewed ethical focus on textual dynamics for research into social systems.

My methodological claim here is that the 'linguistic turn' cited by researchers in communication studies in various fields, while partially informative, must be extended and better focused on textual structures in addition to topical features (content). Such a fuller analysis will enable research to show the internal guides texts construct for meaning and interpretation, the grammatical structures that create selective or preferred methods for interpreting the text, and the pragmatic features that highlight or hide other perspectives and alternatives available to readers. In demonstrating the mechanics of textual work, the linguistic turn can be about more than showing the importance or necessity of texts for social construction, stability and operation. Linking grammatical-level analysis to larger textual features can also demonstrate the mechanics of the rhetoric of change and the textual politics of these key processes.

2 Theoretical intersections: Discourse technologies and autonomous change

> To identify and analyze singularities however is by no means to deny their contemporary status and consequently their interconnections with many other things in the world.
>
> (Paul Rabinow 1999, p. 180)

Proclamations are delivered in unique and ideal rhetorical contexts. As a ritual conjoining of an audience's relative weakness to effect a situation and a writer's unquestioned right to issue a decree, a proclamation appears uninhibited by the whimsical necessities of persuasion. And, so it was, on a small college campus, in a remote corner of the United States during the final days of July 2002. Here, the college's fifteenth president proclaimed a new corporate–university partnership with a multinational computer firm.

This book is about the implementation of this proclamation: the strategic use of a discourse used to explain, compel and drive the partnership; the discursive tensions that occurred as the implementation context became more complex and resistant; and the unique forms of persuasion used to convince people to accept the changes this partnership created. At the same time, this book is concerned with more than just this one relatively minor event. In highlighting this singular example, this study is also concerned with an emerging form of language use associated with the problem of change and more specifically with the implementation of intended and unintended changes within social communities and organizations.

While this book's focus is the implementation of a new software system at a university campus, the singularity of this event occured simultaneously within several different intersections: discourse and change, language and context, universities and corporations, technology and social change. This chapter highlights these intersections, layering them into what, for me, became one multitiered narrative. This chapter begins by continuing the narrative of the corporate expansion into academic sites started in the previous chapter. It describes the specific case which informed this book, the context of this project and the technological changes that provided one narrative structure for this project.

This chapter then turns to a more theoretical narrative to develop an integrated perspective of discourse, technology and persuasion. My claim here is that these three forms have recently been aggregated to articulate a new form of rhetoric within modern organizations. As a new aggregation, this form has appeared as an important enactment of change. The chapter's discussion considers the discursive construction of generative power and the use of strategy, technique and process to disaggregate ethics from discursive activity in modern social contexts. The chapter concludes by examining specific forms of what have come to be called 'discourse technologies', methods of persuasion based on the structure and function of texts.

Corporate universities and academic cultures

As an enactive text, the *corporate university* emerged as a popular and widespread concept in the mid-1980s and continued into 2001–2002 as large corporations centralized their training facilities and branded them 'colleges' and then 'universities'. The corporate university emerged as both a way for companies to standardize and brand their own training and as a new model for higher education that advocates used to impose a new market-based discourse on existing academic university campuses. As in-house training facilities, corporate universities brought together all aspects of an organization's training and human resource development. Hawthorne, Libby and Nash identified 'corporate universities' as centralized training sites created and administered by a corporation whose primary business objective is not education (Hawthorne *et al.* 1983, p. 2).

Corporate universities were tasked with new employee training, continuous education for employees and special courses or programmes that met organization-specific needs. Eurich (1985) predicted that these programmes would grow throughout corporate America and become prominent features not only of corporate life but also of America's higher education landscape. For example, General Motors Institute (now Kettering University) was one of the first corporate colleges in America, dating to a training programme run by Albert Sobey in 1919. Other prominent corporate universities include Motorola University, Harley-Davidson University, Symbol University, Hamburger University (McDonald's Corporation) and the corporate universities administered by General Electric Company's divisions including Crotonville, its company-wide management-training institute.[1]

Although the term 'corporate university' initially referred to centralized training programmes, subsequent arguments and profiles of corporate universities aggregated this 'training programme' discourse within a market-based higher education discourse. In the mid-1990s, pro-market advocates within business and education literatures began claiming that these training centres could represent an alternative, for-profit, market-based model for higher education. In the extreme, the corporate university was cited as a competitor with academic universities. For example, writing about management education, Moore (1997) cited several corporate-run educational programmes that compete with traditional academic MBA and undergraduate business programmes. Moore went so far as to predict that these corporate programmes would force many generalist and less-renowned academic programmes to close. In 1994, Davis and Botkin argued that while traditional universities will continue to be needed to 'carry out basic education and research', companies 'will increasingly take on teaching themselves' (Davis and Botkin 1994, p. 97). They claimed, 'the best' corporate education programmes 'are far superior in their ability to educate than are many lower-tier public and private diploma mills' (p. 101). Six years later, Micklethwait and Wooldridge claimed that there were approximately 1,600 corporate universities in America and they wrote that these corporate centres boasted newer, more expensive and more expansive facilities than most of America's academic universities (Micklethwait and Wooldridge 2000, pp. 68–9). Corporate universities were depicted as newer, better funded and more relevant to the current job market than traditional academic universities.

These arguments ignited a debate about greater market-based initiatives in higher education. Fairclough, a critic of market-based approaches to education, identified what he called the 'metaphorical transfer of the vocabulary of commodities and markets into the educational order of discourse' (Fairclough 1992, p. 208). As I noted in the previous chapter, he demonstrated that market-based educational discourse had created its own vocabulary based on a re-conceptualization of 'skills', 'competence' and the ways these concepts bring together education, career and workplace preparation (Fairclough 1992, p. 209). Fairclough and others showed that the colonization of academic space by market discourse has had tangible consequences on the design of curriculum, the relations between faculty and students, the significance of revenue generation activities and the role of research within academic networks (see, e.g., Aronowitz 2000; Jarvis 2001; Lewis and Smith 1994; Hanna 2000). In addition to these implications for university research and pedagogy, a further consequence of corporate expansion into academic sites is the outsourcing of key campus functions to third-party corporate vendors (bookstores, campus food services, college wear, grounds and maintenance and, as this study documents, information technology).

The corporatizing of technology at Northeastern Tech

One immediate, tangible and significant consequence of corporate interest in higher education became evident at a small technical university on the east coast of the United States, which, for the purpose of this book, I've called Northeastern Tech. The school was founded in the late nineteenth century as a technical college focused on practical and applied training programmes. In these initial years, the college curriculum offered courses to women and men in electrical engineering, domestic science, art, machine work and smithing, woodwork and pattern making and normal manual training. In 1907 the college expanded its engineering offerings to provide degrees in mechanical, civil and chemical engineering. After World War II, the school added graduate programmes in engineering, science and management. Throughout its history the school remained focused on practical, technical programmes, predominantly within engineering disciplines. From 2000–2003, approximately 50 per cent of the total student population was enrolled in engineering programmes with the other 50 per cent distributed across science, business and interdisciplinary programmes, and the arts and humanities.

Located in a largely rural community, Northeastern Tech is one of the area's largest employers, second to the local hospital. Many staff are locally raised and their positions reward them with free tuition at the college for their children. In an economically distressed area, this is a significant advantage as many of the grounds and maintenance workers, administrative assistants and professional staff could otherwise not afford to send their children to college. Like most universities, the faculty is international. The school has a relatively high research expectation for faculty tenure and promotion, especially for faculty in technical disciplines and science. Increasingly, faculty success in obtaining external grants from corporations, foundations or government is having a key influence on tenure and promotion decisions. However, higher administration has been ambiguous about the importance or necessity of peer review for external grants as contacts from prestigious and difficult peer-review processes are seen to be equivalent to contracts for analytical lab services or performance testing of metals, parts or new design specifications.

In the fall of 1999 and into the academic year that turned the century (1999–2000), Northeastern Tech outsourced its Information Technology Office (OIT) to an external, for-profit, corporation. For the purposes of this book, I've called that company Lyceum. A year into this arrangement, Lyceum embarked on a strategic plan which further extended the corporate/university model on campus by negotiating a package with an international high-tech company, which I've called CP Corporation. According to the agreement, CP Corporation would donate to the university project management and communications software, which I've called WorkWare, the servers used to run WorkWare, three years of maintenance, and provide a substantial discount on administrative software. In return, the campus would pay service fees after three years, purchase upgrades and new hardware from CP Corporation, and CP Corporation would become the preferred provider of all non-research personal computers on campus for faculty and staff. Northeastern Tech already had a strong relationship with CP Corporation before this agreement. Two university trustees were high-ranking CP Corporation employees and CP Corporation held a prominent role in the campus career centre, employing graduating students from across all university degree programmes.

Discourse, technology and persuasion

In his 1977 book, *Autonomous Technology*, Langdon Winner proposed that 'technology succeeds *if it is made to appear* autonomous' (Winner 1977, p. 15, emphasis my own). Winner's claim signalled an important and unique pause in the cultural critique of technology. This is not Ellul's (1964) technological determinism, in which human values are subordinated to the efficiency of rational methods. Nor did Winner's pause signal a negotiated, strategic embrace of technology as Price (1965) or Galbraith (1968) had before him. In some ways, Winner's argument was not so much about technology but what we could call the *rhetoric* of technology, the discourse used to situate and promote various technologies within social cultures and institutions.

Winner's project took technology as a starting point and as a way to engage specific processes that enact various types of social change. For him, technologies were both structures and processes. He wrote that technologies were structures 'whose conditions of operation demand the restructuring of their environments' (Winner 1977, p. 100), and processes that 'take on the aspect of self-perpetuation' (Winner 1977, p. 29). What these two characterizations had in common for Winner was the perception that they appeared with 'the force of overwhelming necessity' (Winner 1977, p. 258).

Though he was focused on the intersection of technology and modern society, Winner interpreted much of his topic as a problem of language. A problem, he stated, of insufficient 'resources of public discourse' to provide and enable a meaningful interpretation and understanding of the increasing sphere of technology. Here, Winner agreed with Ellul and others who argued that most non-specialists knew very little about how technologies influence and change their day-to-day lives. But Winner also added another dimension to this critique by arguing that it is not enough to simply know about the technology itself; it is also crucial that we know how the *technology is made to appear overwhelmingly necessary*: how a given technology is constructed and situated as something necessary and crucial to modern life.

Since Winner's book, numerous projects have endeavoured to provide new language for enabling and expanding a cultural critique of technology (see, e.g., Marvin 1988; Zuboff 1988; Postman 1992; Lanham 1993; Cartwright 1995; Latour 1996; Johnson-Eilola 1997; Johnson 1998). In these studies, technology is presented as a material condition and phenomenon. However, as this new language engages, critiques and explains technology, a different sort of problem has emerged. In workplaces, political debates and professional consultations and, as this book will show, in forums of organizational change, language is being deliberately ordered and structured to provide the appearance of rational, instrumental and predetermined imperatives. In such contexts, discourse itself has become a technology as proponents create efficient and strategic means for accomplishing an imperative that requires and perpetuates a specific discursively constructed organization of the world. A *technical discourse* such as this enacts a different sort of persuasion, one not so much based on the message that is stated (the narrative) but the structure and the design of the message itself.[2]

By characterizing certain discourse practices as technologies, one of my primary goals for this book is to connect critical discourse studies with critical studies of technology and sociological studies of change in order to extend and inform an understanding and awareness of the structural processes language can inscribe and enforce in contexts of change. Persuasion, deliberation and argument are key components of social life, vital aspects of a modern democracy and a crucial reason why we learn to use sophisticated forms of language. As Kenneth Burke often repeated, humans are 'symbol using animals' (Burke 1969, p. 3) and persuasion is a long-practised symbolic activity. The act and process of persuasion, Burke writes, is important because 'persuasion involves choice, will'. It is directed to someone only because that person is free to choose among different options (Burke 1969, p. 50).

Given this assumption, that language enables and encourages deliberation and participation, I have used the phrase *the technology of change*, and later *textnologies* to describe two separate but related processes of social change which are at the centre of this project. The first process uses language to construct a world in which participants do not appear to have choices. The second refers to activities in which choice is materially limited but the process of persuasion, argument and deliberation are used to hide this material condition. Both situations enact what Burke called *peithananke*, 'compulsion [enacted] under the guise of persuasion' (Burke 1950, p. 50) and it is this act of discursive compulsion that I am associating with the technology of change. By examining the interplay between discourse and compulsion, between strategies of discourse that situate change as necessary or compulsory, and necessary situations that are presented as choices or options, I want to examine and raise questions about how change is planned, implemented and sustained in social contexts.

When considering technologies of change as forms of social persuasion, we may be realizing a version of what Perelman, in his study *The Realm of Rhetoric*, called 'arguments based on the structure of reality' (Perelman 1982, p. 81). Perelman wrote, 'As soon as elements of reality are associated with each other in a recognized liaison, it is possible to use this liaison as the basis for an argumentation which allows us to pass from what is accepted to what we wish to have accepted' (Perelman 1982, p. 8). Perelman stated that such arguments usually appeal to 'liaisons of succession', which are appeals to cause and effect relationships between two phenomena of the same level (Perelman 1982, p. 8), or 'liaisons of coexistence' which are appeals to relations between unequal levels where one aspect is seen to be a derivative or an expression of the other (Perelman 1982, p. 90). Perelman's study did not consider specific discourse features of these arguments, though his accounting does address how existent views of the world can be leveraged to articulate an opinion or to convince someone to take action. But Perelman's description is not the same as a discourse that operates to structure perception ahead of succession or coexistence.

Unfortunately, Perelman's concept of rhetoric does not have a way to examine language forms that are used to actively construct and normalize a perception of reality or an interpretation of material events. Perelman's account is limited because he does not recognize the constituted/constituting dynamic of discourse nor does he present discourse as something invested with power. Discursive technologies of change expand Perelman's accounting to recognize a field of argumentation based on a perceived autonomous *structuring* of reality as an interplay of discourse, power and technology.

Staying within rhetoric's macro-framework, Rabinow (1999) provided a more constitutive perspective of discourse processes and their ability to co-construct a persuasive reality. He has highlighted Michel Foucault's (1979a) concept of 'biopower' as a key component of the intersection of discourse, power and technology in modern societies. For Foucault, the rise of a normalizing society was central to the shift that provides discourse with its power. Foucault's analysis was historical as he traced the intersection of discourse and power from overt material acts of war and public displays of violence and punishment to modern forms of normalization and social control. He argued that in an authoritarian regime, a sovereign's use of force is intentionally public. Discipline is staged to demonstrate the sovereign's power and ensure civil compliance. Yet, as populations found methods of sustenance apart from feudal relations, people developed a consciousness of themselves as political, moral and civil beings apart from the sovereign. In a post-feudal social structure, the ability to kill, 'the power of life and death' (Foucault 1978, p. 136), became only one form of power among many. As a form of power, public displays of violence became flagrant. Public violence conducted by the sovereign against the populace put the sovereign's legitimacy to reign into doubt. It characterized the sovereign as feeble, as a ruler isolated and removed from the populace.

As a more politically self-conscious population gained realization, other acts of power, intended to 'incite, reinforce, control, monitor, optimize, and organize the forces under it' (Foucault 1978, p. 136), formed what Foucault called a 'generative force' of control. Foucault wrote that these generative forces of power were not designed to impede, make people submit or destroy, but instead, they operated to develop a positive influence on life, to 'administer, optimize, and multiply it' according to specific and precise controls and regulations (Foucault 1978, p. 137). This was not a repressive power, but a productive one. Power that is 'invested' in everyday occurrences. Foucault spoke about 'a power that is not conservative but inventive, a power that possesses within itself the principles of transformation and innovation' (Foucault 2003b, p. 52).

Discourse and technology play two interconnected roles in the formation of generative power. First, the formation of a societal self-consciousness emerged as both a technological and a discursive project. Whereas various technologies enabled a separation from feudalism and an awareness of one's self as a political entity, a self-conscious discourse reflects and constructs this knowledge about oneself and the world. Foucault calls this becoming 'an object of knowledge' which produces both a self-aware discourse and techniques for eliciting and normalizing this discourse. Thus, the eighteenth century's emergence of the professions, and the development of social institutions such as universities, secondary schools and health clinics elicited the recognition of the individual as a new entity and developed techniques to control and place that individual within social communities.

Second, with the decline of discipline as a public spectacle, the generative power to create a stable, motivated and productive citizenry was embedded in a new discourse. These were techniques of power that intersected language and the body operating within education, economics, social institutions (the family, church) and political institutions (army, government, police). Foucault wrote, 'biological existence was reflected in political existence' (Foucault 1979b, p. 142). As forms of power, these discourses were used as techniques to connect biological existence to social and political action. In other words, these new discourses taught people how to live as individual, self-conscious political beings. Foucault wrote:

> the fact of living was no longer an inaccessible substrate that only emerged from time to time, amid the randomness of death and its fatality; part of it passed into knowledge's field of control and power's sphere of intervention. Power would no longer be dealing simply with legal subjects over whom the ultimate dominion was death, but with living beings, and the mastery it would be able to exercise over them would have to be applied at the level of life itself; it was the taking charge of life, more than the threat of death, that gave power its access even to the body.
>
> (Foucault 1979b, pp. 142–3)

Change, in this new context of technical bio-power, does not begin with the act of the sovereign (war, violence). Instead, it occurs in the invention of discourse and the ability of a semi-autonomous politic to use that discourse to report and interpret the activities of the sovereign. Throughout the eighteenth and nineteenth centuries, these new generative discourses of power, which combined the body with ethics and social control, became the domain of emergent professions (Bledstien 1976; Larson 1977;

Friedson 1986). Specifically, they resided in the guilds of law, medicine, the clergy and university teaching and research (Krause 1996). In the period immediately after World War II, generative discourses, which Rabinow identified as discourses of 'life and death, health and disease, pathology and normality', became contested matters of the state (Rabinow 1999, p. 9), critical to issues of security, economic development and national welfare.

More recently, science and capitalism have converged to challenge the state for legitimacy over the control and deployment of generative social power, arguing that private capital and rational science can more efficiently develop, guide and administer modern issues ranging from education and health care to military affairs and environmental and physical resources (Krause 1996; Gee *et al.* 1996).[3] Here, the 'power of life and death' becomes the power to 'administer, optimize, and multiply'. The institution of the sovereign becomes the institutions of the professions, the imposition of the State, and now the rational technique of the efficient capitalist.

It is this convergence of rational capital (scientism) within capitalist business practices that Ellul and others have identified as social technologies, or, the technologization of social life. As Rabinow noted, the convergence of capitalism and science into rational technique has 'disaggregated' the stabilized eighteenth-century fusion of the body, society and ethics (Rabinow 1999, p. 12), placing into contest the ways 'technologies of (social and bodily) recombination are to be aligned with technologies of signification' (Rabinow 1999, p. 12). In other words, whereas the professions, and to a lesser degree the State, generated social power by integrating ethics, the body and society, scientific capitalism has broken these apart and reconstituted generative social power as the technology of rational processes.

The modern fusion of scientific capitalism, with its drive toward rational efficiency and its grounding in objective standards, pulls apart the conflation of ethics, politics and autonomy which created the great eighteenth-century 'individual'. In the realm of technology, ethics and politics are no longer tied to the individual because they are positioned to reside apart from technical issues of strategy, motivation and efficient process. As Steven Katz (1992) and Cezar Ornatowski (1992) have critically shown in separate studies (reviewed below), in a technocratic society, the individual elides ethical accountability to process, since such directives are predetermined by an objective and rationalist calculus of efficiency. Ornatowski has called this a 'deep seated contradiction' between the practical goals of technical discourse and the objective position such discourse claims (Ornatowski 1992, p. 94). Technocratic discourses represent a contradiction for Ornatowski because those who use these discursive forms do not consider (or allow for) the role of the human agent as an interpreter and designer of both the practical goals and the discourse evoked to implement them. The discourse elides any sense of human agency. Simultaneously, Ornatowski argued, the discourse does not overtly admit the priority placed on efficiency or technical prowess. Decisions are made without reference to any guiding criterion whether it be personal (agency) or values-based (efficiency).[4]

As Steven Katz has shown in his analysis of Nazi death camp documents (1992), technologizing processes of 'life and death, health and disease, pathology and normality' generates the appearance that conflicts over values and beliefs can be excised from systems and processes, to be re-articulated in separate spheres removed from the central function of production. Once assumed, the separation of ethics from the sphere of individual action puts into question the ways ethics, values, and even political agency can be re-articulated within discourse. Once discourses of 'life and death, health and disease, pathology and normality' become functions of rational process, how might Rabinow's 'technologies of signification' be assessed for ethical imperatives? What is at stake here is not the reconstitution of the individual as an ethical subject. Rather, it has become vital to locate the ethical space in rational procedural discourse. How may technical discourses of process and strategy be assessed or valued for their construction of the body and the body's own re-articulation within social spaces?

This is not to claim that technologized language is value-free or that it operates in a space apart from ethical judgements. Recalling Winner's assessment, that 'technology succeeds *if it is made to appear* autonomous' (Winner 1977, p. 15), the problem is not the absence of ethical values or the utopia of a value-free process, it is one of articulating where and how these values reside within a framework designed to elide such questioning. In denying the ethical space of language, we are still operating with the belief that language is some sort of neutral arbitrator or hollow shell of transmission. We remain within Winner's central problem of language and technology. But, unlike Winner, our problem concerns a technology that has come to deny its discursivity and a co-constituted technical discourse that denies its ethical framework. In the optimization of process, language is made to appear autonomous from ethics.

Discourse technologies and social change

In practical terms, most people will recognize (typically after the fact) that certain discursive strategies, for example, using the passive voice to elide agency ('decisions are made without reference to any guiding criteria'), treating new information as given, creating grammatically coherent links between dissimilar concepts or constructing an ultimatum to achieve consensus all present an audience with false choices. Recalling Jobs' speech as presented by Harvey in the previous chapter, embedded within the false choice between embracing change or inviting defeat, between those who would succeed and those who would fail, those who would continue to work and those who would not, lies the discursive enactment of Winner's autonomous technology. The above-reported ways of constructing strategic ambiguity technologize choice, consent and participation. Although these actions happen at the grammatical (discursive) level, they present audience's support for organizational leadership and acquiescence to planned change as rational and deliberate acts, even though these audiences in actuality have little, if any, choice in controlling or directing the social changes around them.

This chapter has examined the separation of ethical accountability from specific discursive practices, claiming that with the rise of modernity, language use has been situated to appear autonomous from ethics. In this context, how a message is delivered/written has become separated from both what that message is (what is said) and the assigning of accountability for how that message is designed and delivered (ethics). I introduced this claim by referencing Kenneth Burke's concept of *peithananke* which he described as 'compulsion [enacted] under the guise of persuasion' (Burke 1950, p. 50) and Chaim Perelman's description of 'arguments based on the structure of reality' (Perelman 1982, p. 81). Together, these concepts describe functional language-in-use in which discourse itself becomes a technology – a way to perpetuate a structuring of reality.

When used as a technology, discourse enables advocates to create efficient and strategic means for accomplishing imperatives without requiring a consideration of the means by which this takes place. Such a constitutive form of rhetoric casts Winner's suspicion, that technology becomes successful when it is made to appear autonomous (Winner 1977, p. 15), as a discursive problem – a problem of communication design and use. Thus, in contexts of social change we can see language being deliberately ordered and structured to provide the appearance of rational and instrumental yet predetermined imperatives. Foucault characterized such discourse as a power that is inventive and constructive; yet, it is removed from personal (or institutional) accountability or the need (desire) for explication (Foucault 2003b, p. 52). Jon Simons (1995) added that such a combination of 'technical and purposive rationality' in discourse constructs a language that appears, in his words, 'programmatic'. Such language is intended to induce 'methodically rational conduct which is simultaneously rule-governed, reproducible, and principled' (Simons 1995, p. 38).

Whereas the preceding discussion examined the emergence, function and influence of technocratic discourse, I will conclude this current section by more finely elaborating the features of technocratic discourse as they have been presented in discourse studies literatures. As an ethical discussion, the perspectives brought together here represent a broad consensus among discourse scholarship and my purpose here is not to dispute these judgements. These accounts diagnose a discursive practice/problem in which specific forms of language are used to compel persuasion by constructing reality in ways biased toward the interests of those who benefit from specific social changes.

Yet, as a way to forecast the chapters that follow, my contention will be to suggest that in a context of social change the use of technocratic discourse is a normative in addition to an ethical problem. Let me explain. In changing stochastic social systems, discourse functions to restore and stabilize dysfunctional (destabilized) networks. This process of restabilization, to varying degrees, must be competitive and agonistic as various agents (or collective groups of agents) come together to pursue courses of action that reflect their best interests. A central feature of this book, demonstrated in the following chapters, is to explore the features of discourse that enable such a process of change to occur. As discourse stabilizes social contexts into new/changed social orders, it will take on formal and functional characteristics that suggest restabilization and the creation of a stable order.

My suspicion is that a technologized discourse enables speakers and writers to initiate this period of restabilization because at this moment there is no foundation to reference, no 'actual' to invoke, yet there remain competing claims and representations of what will occur. In other words, discourse technologies promote action and acceptance (belief) in that moment between regimes, between casting off the known and accepting the unknown. This is both a descriptive claim and an ethical claim. In the absence of actual material evidence (the change has not yet happened) advocates must find a compelling reason for participants to embrace the new. But, technological efficacy does not mean that a technologized discourse should continue to be held apart from ethics as modern distinctions between 'form' and 'content' would claim. Nor does this mean that discourse technologies are the only way in which to bridge the gap of instability during change. Instead, it means that critical studies of change can conceptualize an examination of the 'ethics of change' which includes a critical accounting of the process of change and the ways language has been used to create and stabilize new regimes – in addition to the more typical critique of the 'content' of change messages themselves. This chapter's concluding discussion is positioned then as a transition from conceptualizing an ethics of technologized discourse to understanding technologized discourse as a (potentially normative) feature of changing social systems. This is not to leave questions of ethics and accountability behind; it is a way to question how the role and function of discourses of social change may inform and potentially alter this ethical framework as it has been already formulated.

To account the more granular text-level features of technologized discourse, I have summarized Jay Lemke's (1995) model of technocratic discourse. Lemke's account is both descriptive and critical and although he does not overtly trace processes of systematic social change, his report emerges from initiatives to create value-oriented social policies, especially in education. I have correlated Lemke's findings with Norman Fairclough's model of *technologization of discourse* (1992, pp. 215–18, 1995, pp. 102–11, 1996) which articulates the use of technologized discourse within newly emerging social contexts of work, especially in organizational and professional contexts. The chapter closes by reviewing Ornatowski and Katz's separate work locating technocratic efficiency within the academic field and corporate practice of technical communication.

Jay Lemke: Technical technocratic ideology

Lemke has characterized the function of technocratic discourse as 'the transformation of discourses of expert knowledge into discourses of social policy' (Lemke 1995, p. 58). Lemke wrote that the advantage of this '*technocratic* strategy' (Lemke 1995, p. 58, emphasis in original) is that it allows speakers and writers to make policy claims appear as if they were created exclusively from facts. These strategies have the effect of 'deflecting' the various choices of values that determined the policy and 'the social, moral, and political responsibility for such choices' (Lemke 1995, p. 58). Lemke described both the specific technical features of this discourse and its social function as an instrument of persuasion and policy creation. Describing the specific textual features of technocratic discourse, Lemke wrote that such texts (or talk) express few processes of direct action. Instead, they accentuate abstract relations between agents and activities (Lemke 1995, p. 60). He noted that such texts are highly nominalized. Processes are described as participants and animate relations between people and people, or people and ideas, or people and institutions are elided and rarely appear. Rather than detail agency, writers will use agentless passive clause structures that can grammatically take the place of human agents.

Table 2.1 Lemke's technical discourse summarized

'accentuate abstract relations between agents and activities'	
Feature	*Discourse Characteristic*
Animate relations elided	Agentless passive clause structures
No forum for discussion	Limited pronoun use (I, we), monological voice
Arguments, decisions condensed	Arguments fronted by desired actions with no details about how argument was formed
applying apparently scientific rules and certainty to ideological and social issues	

In addition to using grammatical features that elide agency and accountability, such texts are dominated by third-person forms (Lemke 1995, p. 60). Writers will rarely use 'I' or an inclusive 'we' (referring to the reader and writer together). As a result, no space is allowed to elaborate, encourage or admit a process of dialogue or disagreement. Citing

Bakhtin's (1986) description of the many voices and opinions that often come together to constitute texts, Lemke wrote that technocratic texts uniquely appear 'monological'. In other words, these texts appear to be formed by one voice and opinion and they situate their content as something that is not the result of debate or discussion but simply a single known 'fact'.

As a way to link the formal characteristics of these texts with their social function, Lemke used the term 'condensation' (Lemke 1995, p. 60). Formally, the term refers to the grammatical features which reduce (condense) agency and voice (largely through nominalizations, passive clause structures and pronoun elision). Functionally, Lemke argued that such discourse reduces or eliminates audience participation in policy decisions. He wrote that such texts 'incorporate condensations from technical discourse without requiring that the condensations be expanded in order for the text to be meaningful to the reader' (Lemke 1995, p. 71). Technocratic discourse allows writers/speakers to gloss over the technical aspects of an argument (or even negate or reverse them) since the genre does not obligate the writer/speaker to alert the audience to the reductions that lie behind the argument (Lemke 1995, p. 75). Such language appears solely as a value-neutral, objective reporting of 'the facts' (Lemke 1995, pp. 60–1). It is not intended to be presented or used as 'just one more opinion' or qualified assumption about an issue. Instead, the discourse is situated 'above the fray' (p. 70), as a neutral and objective account of facts. Lemke wrote that such a position is rhetorically powerful, because the text claims to be 'free of all interests and values except truth, which *all* parties *must* take into account in deciding policy' (Lemke 1995, p. 70).

For Lemke, the difficulty with such technocratic texts is that they apply apparently scientific rules and certainty to ideological and social issues. Lemke writes that 'in every case' the discourse is initiated with 'an action it covertly wants to recommend as policy' (Lemke 1995, p. 71). This action is then supported by research evidence, studies and findings 'which show that this action is a necessary cause of something else that is positively valued by those to be convinced' (Lemke 1995, p. 71). The means of getting to the action, the values entailed by the action or the outcomes of the action are not included or considered in the argument. Instead, Lemke claims, 'the argument is only that the covertly recommended means are necessary to the unargued ends' (Lemke 1995, p. 71).

Norman Fairclough: Technologization of discourse

A second conceptualization of *discourse technologies* was elaborated by Norman Fairclough's use of the term *technologization of discourse* (Fairclough 1992, pp. 215–18, 1995, pp. 102–11, 1996). Like Burke's use of the term *peithananke,* Fairclough used technologization of discourse to refer to specific forms of language that are consciously designed to create strategic, efficient and rational solutions to problems (Fairclough 1992, p. 216). While Lemke's description largely applied to the use of technocratic discourse employed to address social and ideological problems, Fairclough's use extended the range of issues to include a larger scope of human persuasion including human resource activities within organizations and professional consultations between doctors and patients or other such encounters. In addition, Fairclough added the additional claim that such discourses have emerged from and are grounded in forms of social science research and experimental design (e.g. usability research, audience analysis, focus group research, sociolinguistics) and they are often products of institutional and professional training and education (Fairclough 1996, p. 71). For example, Fairclough cited the use of interviews in human resource practices in which those conducting the interview (those in power) consciously create an informal atmosphere, appear to grant the interviewee equal status, and enact specific discursive strategies in order to persuade the interviewee to reveal private information (Fairclough 1992, p. 218).

Whereas Lemke's account diagnosed formal and functional features of technologized discourses, Fairclough outlined the conditions that discourse technologies constitute as a product of the specific textual features of enacted discourse. Certainly, Lemke's description informs and elaborates at the textual level the accounts Fairclough provided and can be used as a way to elaborate the specific discourse practices Fairclough detailed. Yet, Fairclough's elaboration stipulated a more intentional process of creation, review and testing that placed technologized discourses within a system of discourse design and use. Fairclough articulated five key features of technologized discourses: they are developed by social science research; checked and refined for results; produce new genres; evoke solidarity with their audience; and promote standardization and normalization across various sectors (Fairclough 1996, pp. 73–5).

First, Fairclough wrote that discourse technologies are created and used by social scientists and other experts in communication and language studies. This association with professional and scientific status provides those who use such discursive formulations with a degree of prestige and the appearance of expert status. In addition, the discourse formulations are based on scientific research and study. They have emerged as an applied research product designed to effectively influence and persuade. However, as noted above, though research has shown specific technical features that influence persuasion, a critical discussion about the appropriate use and value of these techniques has not accompanied this empirical work. Second, Fairclough argued that discourse technologies are continually checked and corrected to maximize audience reception. This revision is conducted by outside experts who advise new genres, frameworks and discourse patterns. For example, a speech writer will continually revise a politician's public statements as situations, audiences and other aspects of the discursive situation change.

Third, discourse technologies appear in specific and new genres that are consciously designed to influence an audience. Fairclough highlighted interviews, training sessions, managerial discourse and marketing discourse as genres reconstituted as technologies. Fourth, these genres are often hybrids that combine formal and informal contexts and institutional and private discourses to appear democratic, personal and evoke a degree of solidarity within public organizational contexts. Fifth, Fairclough wrote that discourse technologies promote a 'standardization and normalization' of language practices across institutional and organizational contexts. Such practices establish a commonality of discourse across various economic sectors and have been associated with new globalized capitalism (see, e.g., Gee *et al.* 1996; Micklethwait and Wooldridge 2000).

Cezar Ornatowski: Technocratic style

A final characterization of technocratic discourse has remained a subject of some controversy in the applied academic field of technical communication. Researchers in technical communication have long been suspicious of overly positivist textbook accounts of technical communication, which, as Ornatowski has written, 'locate the essence of technical writing in objectivity, clarity, and neutrality, and the need for teaching it in its usefulness to employers' (Ornatowski 1992, p. 91). Steven Katz (1992) has gone so far as to argue that the primary value of the field has become 'expediency'. He wrote that 'an ethic of expediency . . . underlies technical writing and rhetoric' because most technical communication is deliberative and in modern society most deliberative discourse is technical (Katz 1992, p. 273). Judging from the continued reactions to his argument, Katz's mea culpa for the field was memorable but also seen to be extreme. Though he rightly demonstrated the problems with a value system dedicated entirely to maximizing efficiency, his broad strokes and generalizations were problematic for a large and diverse field of researchers, teachers and practitioners who are informed by numerous issues, interests, demands, values and expectations.

In contrast, Ornatowski (1992) offered a similar yet more nuanced critique of the role of technocratic discourse in the field. Ornatowski treated technocratic writing as a stylistic choice for technical communicators, one 'rhetorical tool' among many. Technical writers, claimed Ornatowski, 'are rhetoricians who continually make ethical choices in serving diverse interests and negotiating between conflicting demands' (Ornatowski 1992, p. 92). Recognizing 'the fundamental rhetoricity' of technical communication, he noted, introduces choice, accountability and ultimately ethics back into the the activity of technical writing. The concept of choice and the potential of 'going technical', to put the phrase into a vernacular, seems to better account for what Ornatowski called 'the complex nature of the relationship between language and its social contexts' (Ornatowski 1992, p. 93). From this perspective, using the passive tense, inserting nominalized processes in place of agents and hedging possibility through modality become stylistic devices for which writers can be held accountable.

In concluding with Ornatowski's option to invoke technical discourse, we have returned (again) to ethics, though unlike Fairclough's notion of a surreptitious technocratic discourse, Ornatowski's technology is employed overtly and by choice. Ornatowski overtly realized the utility of his technical writing: 'it is a vital element in furthering the goals (business, political, economic, and other goals) of organizations and institutions' he wrote (Ornatowski 1992, p. 102). Further, he stressed,

> the essence of its usefulness lies in the fact that it is capable of giving a rational, technical, "scientific" mantle to agendas that are always in some measure interpretations. The fact that we do not commonly think that way about it is its greatest rhetorical asset. Realizing this is the first step towards having the capacity to analyze the trade-offs and bargains that one makes. And that is the first step to talking sensibly about ethics.
>
> (Ornatowski 1992, p. 102)

For Ornatowski, a writer's choice to 'go technical' represents a strategic compromise that places the exigency of a situation in potential contrast with the agency and values of a writer. It is a drama that relies heavily and equally on informed and perceptive writers and audiences for resolution. But it does not fully resolve our problem with the functional responsibilities of a discourse that enacts change nor does it tell us much more about the necessary (and tangential) functions of language during change. Does assigning the functions of a change discourse to simple rhetorical choice signal a capitulation to a sort of individualist relativism – that people will do what they think best at the time? Or, as I speculated above, does technical discourse evoke something more foundational and necessary in the operations of language during the freefall of change? These are the questions that frame the study in the following chapters and they are the questions we can use to transition to the context of the WorkWare narrative I will be exploring in these chapters.

In examining the function of written communication in contexts of change, researchers have stressed the important function of written discourse in influencing cognition and perception within social systems. This research has shown that social change occurs as a combination of material and discursive events. The social realm is always mediated by this interplay. But, recognizing a role for material and contextual elements in the process of change should not lead researchers to discount the role of language and the symbolic in these same processes. In fact, when language becomes strategically and structurally deployed to attain specific goals, the distinction between discourse and technology begins to blur. Typically, we could argue that while material conditions may initiate disruptions in social systems, discursive activity in the form of symbolic activity, written communication and speech provides the necessary means for perceiving, interpreting and acting on the material disruption. Yet, as we will see in the chapters that follow, language itself can become a material agent that disrupts a social system. As such, understanding the function of strategic symbolic activity, as technocratic, deliberate, poetic or narrative, remains crucial to understanding the machinery of change within social networks.

Discussing the evolution and role fads play in current management practices, Micklethwait and Wooldridge were especially critical about inflated claims regarding corporate learning and the knowledge worker. They wrote that training departments are 'forever on the lookout for ideas that will help inflate the training budget, while the "human resource specialists" are suckers for any theorist who argues for "putting people first"' (Micklethwait and Wooldridge 1996, p. 62). Micklethwait and Wooldridge's dismissive may be appropriate. Perhaps corporate universities have been a creative way to leverage more resources for trainers and training budgets.[5] Yet, this account does not explain the impact and longevity of fast-capitalist discourses that accompanied the rhetoric of the corporate university as training centre. Thompson summarized this irony well, noting that the mid-1980s prophecies of 'dramatic growth and significance of corporate colleges remain[s] unfulfilled although ... the broader corporate involvement in education and training in the United States and Canada remains enormously significant' (Thompson 2000, p. 340).

In other words, although the model has provided few results and little success within its own corporate constituency, the ideology of the corporate university continues to influence and change academic higher education. Here, then, is the discursive 'sleight of hand' I hope this book documents: how does it happen that a failed ideological movement (or in this case a semi-failed technological endeavour) can achieve significant and important social changes without ever overtly making the semantic case for these same changes?

3 Discourse/technology: Heterogeneous narratives

> The more a technological project progresses, the more the role of technology decreases, in relative terms.
>
> (Bruno Latour 1996, p. 126)

> of course, in life we never begin at the beginning
>
> (Lorenzo Simpson 1995, p. 60)

For people who study semiotics and organizational symbolism, the emergence of new technologies remains a considerable empirical and social problem. Take, for example, Latour's study of a new transportation system in Paris (1996). The actual emergence and eventual stabilization of such a technology has as much to do with social values, personal characteristics and preferences and social trends as it does with effective machinery and sound design. Goodall accounted for the decisions that influence a person's choice of technology when he wrote, 'in this culture of hypercapitalism and commodity values you are what you drive at least as much as you are what you wear, eat, listen to, or talk about' (1994, p. 136). Symbolically, a person's transportation choice, either subway, automobile, pick-up or bicycle, is tied to a sense of personal and professional identity and self image, as controversial as those terms may be. For others though, the choice of vehicle is nothing more than a mere selection of appropriate and functional technology. Scaffolding the dynamic between these two positions is modernity's interpretive duality between discourse (symbolism) on the one hand and technology on the other.

This well-established duality represents a relatively conventional perspective that positions forms of symbolic mediation as antagonistic and separate from technological mediation. In its extreme case, proponents of technology embrace technological mediation because they believe technology is not discursive; because solving problems through a machine or computer interface apparently does not require textual or human interaction (Krug 2005, p. 21).[1] At the other extreme, cultural theorists argue that discursive mediation in either text, speech or graphic design is creative and fully interpersonal. At the extreme humanist position, meaning is jointly or even multiply constructed in an ongoing interplay among human agents in a context that is free from (or not constrained by) technology's interference.

The purpose of this chapter is to pull these admittedly exaggerated constructions more closely together and in doing so simultaneously pull apart assumptions about technology and discourse. The chapter interweaves the narrative of the WorkWare implementation with a theoretical examination of discourse/technology. Rhetorically, this presentation fulfils two related goals: first, the descriptive narrative provides a grounded context for the more analytical chapters that follow; second, the theoretical narrative elaborates what it means to claim that technology and discourse are mutually constitutive – that technology is a form of discourse and that discourse can be considered a form of technology.

That said, the trouble with structuring discourse/technology as a simple duality is that such a construct does not admit or enable what is a more complex structuration. Chouliaraki and Fairclough (1999) argued that 'the semiotic constitution by the social and of the social is constantly at issue in language analysis' (p. 140) and the constitution of technology is no different. Their caution that we must 'recognize the social import of discourse without reducing social life to discourse' (p. 6) reminds us that viewing discourse as a constructive component of technology does not reduce the constitutive role of discourse. But at the same time, neither does it give permission to reject other components that make up a technological field. Thus, it is important not to 'overdetermine' (p. 6) the constitutive function of discourse in social contexts but to also recognize other factors that contribute to the emergence and ready conceptualization of discourse/technology.

Constituting the field of social change

Elaborating the dynamics that enable the creation of a social field, Harvey (1996; also in Chouliaraki and Fairclough 1999) proposed six fields that he argued construct social processes: discourse/language; power; social relations; material practices; institutions/rituals; and beliefs/values/desires (Chouliaraki and Fairclough 1999, p. 6). Each of the six fields encompasses and realizes the others, creating a 'heterogeneous' moment of instantiation and social creation. Harvey's framework suggested that discourse studies can characterize technology as constructed and mediated not only by discourse then, but also by power, social relations, material practices, institutions/rituals and beliefs/values/desires while simultaneously recognizing the interaction of discourse within each of these other fields. Although the overt interest of this chapter is the overlaying of discourse and technology, to demonstrate the heterogeneous scaffolding of these concepts, the WorkWare narrative and the accompanying theoretical discussion will be structured by each of Harvey's six fields. This is not to say that these fields are exclusive or obligatory, instead, they represent a useful and informative characterization of the social and theoretical fields that make up the technology we are calling WorkWare.

Institutions/Rituals

From: Martin <martin@university.edu>
To: University_Campus
Date: Mon 29 Jul 2002 18:26:13 -0400
Subject: WorkWare

As President [...] mentioned in his correspondence to the campus on 24 July [Northeastern Tech] has entered a strategic partnership with the CP Corporation. One of the many benefits of this partnership is the implementation of the WorkWare application.

The purpose of this communication is to present the key issues associated with the implementation and to ask you not to worry if you don't understand all of the terms you will hear over the next few months. You will be receiving several email communications in the coming weeks that will give more definition to the new application. If you remember only one thing from this letter please let it be that the Office of Information Technology is committed to helping you use this tool to assist in accomplishing your daily activities.

What is WorkWare? It is much more than just a new email system. In addition to traditional email and messaging, WorkWare will enable integrated calendaring, document management and work flow, instant messaging and collaboration software and an electronic workspace. Rest assured, as these services are introduced, plenty of hands-on training and user forums will be available. The end result of this endeavour is to provide you with the latest technology to help you work more efficiently and ultimately provide our students with the best educational and research opportunities.

This exciting project is divided into two phases, Phase I being the WorkWare roll-out and Phase II being the incremental implementation of the full WorkWare suite of collaboration tools. Phase I, the introduction of the WorkWare client, is scheduled to begin on 15 August 2002. WorkWare is a full-featured product that incorporates email, calendaring, group scheduling and information management. As more features of the WorkWare suite are added, WorkWare will be able to be used for document sharing, work flow management and establishing an electronic workplace that makes your documents available anywhere you have Internet access.

The WorkWare client is an Internet interface that gives you the most of the features of the WorkWare client, and is available to any Web browser. With WorkWare, whether you use Windows, Linux, Solaris or Macintosh, you will have access to your email and calendar, anywhere and anytime you have access to the Internet.

Orchestrated efforts for the smooth migration of current email, user training and demonstration sessions are being formulated by the Office of Information Technology and the WorkWare Implementation Team. With this new environment, we will be moving towards a standard 1+7 naming convention, so your user name will be consistent across the many systems we use at [Northeastern Tech].

In conjunction with the user name the new standard for email addresses will also be 1+7. For example, Ben Franklin would be bfrankli@ ... We are fully aware that many people have email addresses that they do not want to change and OIT is more than willing to accommodate this. We will keep you well informed as to when these changes occur.

We realize that this announcement will generate many questions. To address these questions, and demonstrate the many features of this new software, we will conduct two open forums. We invite the campus to join us on Wednesday 31 July or on Thursday 1 August, at 1:00 pm in [location]. There is also a very informative Web page that explains the project in more detail. To become more familiar with WorkWare please visit www.university.edu/~workware.

As this project progresses, this Web page will be updated with current information and hints. If you have specific questions, please feel free to call Abigail Ridley the Project Manager for this implementation, at x6789, or you can email the project team at WorkWare@university.edu.

Figure 3.1 WorkWare implementation email 1

Of all the new technologies to appear in modern workplaces over the last two decades, few have permeated quite so completely as electronic mail. Email's easy infusion into modern organizations has enabled and accentuated well-known challenges (both old and new) for modern organizations. Yet perhaps the technology's greatest consequence has been in the way many people now take for granted the ability to send and receive near-instant text, pictures and video (and numerous other things) to a now international in-box. As a result, perhaps few pieces of modern workplace technology are perceived to be quite as personal as email. Add instant messenger, document management and collaboration and scheduling to the package and it becomes evident that in one proclamation, the intended plan would replace and impose the one piece of software most people daily use to construct and represent themselves to rest of the world. To put it in more pragmatic terms, given the scope of modern academic organizations, overhauling email threatened to further complicate the business and academic relationships university employees had cast across the world. Although this arrangement, and the university president's decree which announced it, appeared to manifest something relatively simple, a mere shift from one piece of technology to another, as usual, the human dynamic influenced by the change belied the assumed simplicity of the deal.

The scaffolding for this, the first campus-wide email initiating the organization-wide change to WorkWare, was constructed during a textbook four-month consultation period. But these consultations occurred within an intensely problematic social context. Martin, the college's Chief Technology Officer (CTO), was the third CTO in five years and his presence enabled a certain amount of stability within an uneasy and suspicious office. It would be difficult to over-state the connections between Martin's personal ethos, his new department's reflection of itself, and its credible image on campus. He was brought in to repair the deep structural damage incited by two years of restructuring and downsizing – a narrative that has now become clichéd throughout modern organizations. His immediate predecessor was hired to dismantle a computer support centre that had become solipsistic. Her strategy repealed all of the technical positions within the office and re-advertised them within a newly structured unit, Computer Information Systems or CIS.

The resulting purge created a significant distraction within the office with two key results. First, the campus's technology infrastructure was left to deteriorate while CIS sorted out its human resources crisis. Second, most campus computer-users lost what little confidence they had in the model of a centralized computer support service. Each academic and administrative unit was left on its own to purchase machines, operate software and create electronic networks for printing, email, scheduling and Web hosting. Several professors even went so far as to set up their own servers with websites and email accounts for other professors independent from the deteriorating and neglected main campus network. Two years into this restructuring, when the technological and social infrastructures within CIS finally crumbled, the university's president declared the situation untenable and sub-contracted out the entire service.

This, then, was how, 17 months before the WorkWare implementation was launched, Martin arrived on the Northeastern Tech campus as the director of Lyceum, a third-party, for-profit company the university contracted to provide computer support services on campus. Although he was not a university employee, Martin was given a cabinet-level position within the university administration equivalent to a vice president. Those employees who had survived the recent office purge could choose to work for Lyceum or for the university; the nature of their day-to-day employment would not change. The majority of salaried employees chose to work for Lyceum and every new management hire within the department was made as a Lyceum employee.

One of Martin's first initiatives as the director of the newly-constituted Office of Information Technology (OIT) was to initiate a series of campus-wide consultations to evaluate the current Information Technology (IT) infrastructure. The result of these sessions, he claims, determined that a consolidation of services had to become his office's first priority. Most pressing, he concluded, was the need for a common communications and information platform to integrate email, scheduling and other administrative functions. Such a feature, ideally, would clean up the variety of discordant software applications, platforms and individually built rogue programs being used across campus. In their place, Martin's office would install and support one campus-wide application. Given the historical animosity IT services faced on campus, Martin did not want to appear to be forcing a new product on the campus community. He created an advisory group of 25 campus representatives who were tasked with evaluating three alternative products. His staff attended advisory group meetings and acted much like legislative aids to the committee members. The result of this process was a White Paper and decision-making rubric which identified WorkWare as the most appropriate solution for the campus. As Martin recalled, anticipating a then unknown future, he wanted to be sure that he could demonstrate that WorkWare was not an OIT decision, but was recommended for its pricing, functionality and other features identified by the campus advisory group. That said, Martin was fortunate that his advisory group recommended a product developed by a key corporate partner of the university (CP). But, in retrospect, he was equally unfortunate in that his campus would be CP's first institution-wide implementation of WorkWare.

Institutions, rituals, and social change

Institutions and rituals remain a problem for organizational change. Change efforts are typically designed to replace deeply embedded institutional practices (rituals). But, more problematically, embedded practices structure ways of perceiving that make it difficult, if not impossible, to construct a viable and agreed representation of empirical experience within a social context. Simpson put this dilemma another way, arguing that 'attempts to provide a "clean" antiseptically disingenuous account of a form of life are ineluctably "sullied" by the inescapability of taken-for-granted conventions of description, canons of rationality, criteria of significance, and sameness and difference, and so on' (1995, p. 112). All positions of representation and interpretation are necessarily mediated by any number of forces and perspectives.

It is worth noting that Simpson used a discourse-based metaphor to elaborate his claim, writing, 'like the search for a language that effaces itself as a language, the attempt to occupy the pure standpoint of the observer, where no unexamined directives for seeing things this way or that are taken for granted, must end in failure' (Simpson 1995, p. 112). If this suspicion of a pure standpoint for observation is well known in social science and humanities research, the concept of a disinterested agent remains attractive for technological mediation. Discourse technologies (see discussion in Chapter 2) are especially problematic because they evoke the connotation of discursive tools, techniques and resources that, though wielded in the support of special interests, have in themselves the appearance of objective or rational refinements. Fairclough (1992, p. 215) suggested that these forms of communication 'have taken on ... the character of transcontextual techniques' that appear as 'resources or toolkits'. In other words, the concept of discourse technologies seems to hold out the possibility of a neutral discursive tool, a language that effaces itself as a language. One may use the interview to achieve a specific ideological goal, but the interview itself is not seen as especially laden with ideology or specific value.

Yet, interviews, teaching, counselling, conversation skills and other discourse technologies are equally products of institutional contexts and rituals. Using Latour's (1996) vocabulary, discourse technologies can be seen as *assemblies* or *assemblages* of institutional and ritual practices. He wrote that in the deployment of these assemblages, the 'form, scope, and power' of institutional context will continually change, introducing new and removing used forces from the operating assemblage (Latour 1996, p. 154). This is why he suggested replacing the term 'context' with 'network'. Since, he argued, a technological project 'is not in a context; it gives itself a context, or sometimes does not give itself one' (Latour 1996, p. 133).[2] What Latour meant here is that technologies are not deployed within an isolated and pure environment, Simpson's 'antiseptically disingenuous account'. Instead, the aggregation that constitutes a technology enables and specifies the conditions for its use. Technologies are deployed within an aggregation of contesting forces that wish to shape and own the technology for their own purposes.

Preferring the term 'network' over 'context' articulates the multiple influences that co-construct the discourse technologies deployed at that moment. Simpson made a similar argument, writing that all the factors which influence a technology 'must be transformed in accordance with the necessities of that complex' (Simpson 1995, pp. 21–2). Simpson was referring here to an interdependence among institutions and rituals which work together to construct a specific technology. He provided the example of automotive technology which is dependent upon but also has enabled and managed developments and expectations in a network that includes the petroleum industry, road building, steel, plastics, computer software, consumer desire, aesthetics, architecture and time management (Simpson 1995, pp. 21–2). These various institutions and rituals aggregate together to constitute an institution called *automotive technology*.

This work of contextualizing makes it difficult to predict or anticipate the specific fields that will constitute a specific institutional network for any given technology (Latour 1996, p. 137). 'The problem is,' wrote Latour (1996, pp. 56–7), 'the innovation has to count on assemblages of things that often have the same uncertain nature as groups of people'. In other words, a technology does not stand on its own, but must interest other fields in order to aggregate itself (constitute its network). Not only is it difficult to predict the different fields that will constitute a network and ultimately a technological institution, but within this aggregation, interpretations of success often differ across the different groups that make up the network (Latour 1996, p. 67). In Latour's words, a successful technology then, 'must at once recruit new allies and at the same time make sure that their recruitment is assured' (Latour 1996, p. 71). Recruitment is promoted through what Latour metaphorically called 'little bridges' that temporarily exist between the technology and the network/institution (Latour 1996, p. 175).

At this point, we have moved far away from the rational and objective antiseptic ideal of technological mediation. Rationality is supplanted by seemingly random assemblages of self-interest and dependence. Neutrality has been traded for declarations of support and preservation. But, this is not to suggest that the technology is in some way compromised or ineffective. Rather, the technology works precisely because it has been institutionalized. It works because of the institutional network aggregated around it with the intent to promote and support its material effects. The institutionalization of technology evokes Latour's 'paradox of development' which I used to introduce this chapter, 'the more a technological project progresses, the more the role of technology decreases, in relative terms' (Latour 1996, p. 126). The materiality of a specific project becomes less important as the institutional network embeds itself within it.

The network space opened by the declining necessity of technology's materiality necessitates an institutional rearticulation: a rhetorical performance accounting for the specific technology. Latour called this account a 'supple frame of reference', writing that, 'if the actors in a project define not only the other actors' essences and desires but also the rules for interpretation that make these definitions workable, all their viewpoints have to be deployed in a supple frame of reference' (Latour 1996, p. 170). But how are these frames of reference agreed, distributed and deployed? Some accounts view these frames deterministically as deployed forms of power. Krug, for example, argued that 'each piece of technology, together with technology as a system, works with other beliefs toward creating a unified totality that does not admit for variance' (Krug 2005, p. 19). Yet, Krug's perspective elides the function of these 'other beliefs' which influence or even predetermine the assembled account within this apparent totality.

Johnson-Eilola provided a more nuanced description of these institutional accounts, acknowledging that 'rearticulations frequently represent a chicken-and-egg problem' (2005, p. 41). Using the emergence of new high-tech consulting careers as an example, he noted that institutional aggregates often leave us wondering: 'Which came first? symbolic analytic workers or interfaces and social contexts to support them?' Johnson-Eilola argued neither, suggesting instead that 'each emerged slowly, contingently, gathering force from the other (as well as other mutually reinforcing forces)' (2005, p. 41). Or, as Stuart Hall put it, 'unity' (of discourse, technology, institution) 'is really the articulation of different distinct elements which can be rearticulated in different ways because they have no necessary "belongingness"' (Hall 1989, p. 141).

Technology does not appear within an organization or an institution autonomously preformed and neutrally aligned to achieve a specific materiality. But, like the fictional language that effaces itself as a language, technology desires to efface itself as an ideology. To do this, technologies become aggregated within an institution and within specific social networks. Their advocates and creators draw upon relevant power structures and discretely embed their work within specific ideologies and habitual practices. Finally, a technology must become aligned with and therefore must articulate a specific institutional discourse which affirms its utility while providing a self-effacing rationale, a retrospective continuity for its implementation.

Institutionalizing WorkWare

The WorkWare recommendation was distributed in a White Paper titled 'Groupware Solution Project Initiation – Analysis and Recommendation' and co-authored by Abigail Ridley (the designated point person in Email 1) and a junior assistant. It was distributed in August 2001, eight months after Martin arrived on campus. The document itself established a simple problem-solution narrative citing the use of 'heterogeneous clients', 'diverse environment' and 'concerns for our client base and CIS support staff' as issues requiring resolution. In a relevant and prescient gesture, the document situated these as community issues, arguing:

> Real-time collaboration is becoming a critical communication tool, enabling virtual teams to meet and work together across campus and beyond campus boundaries. It is essential that our academic and administrative environments have an automated tool to integrate communication into our pedagogy. After all, '*Everything we do here begins and ends with students*'. [Northeastern's] groupware selection should ultimately bring together three crucial communication tools: email, calendaring and real-time collaboration.

The awkwardly-placed statement 'everything we do here begins and ends with students' aggregated into the project a marketing slogan then perpetuated by the university's president and senior administration. The rhetorical logic of its placement here is not so much to document any specific functional or technical claim about WorkWare but to aggregate the White Paper and the WorkWare campaign within the vernacular of the current campus leadership.

In many ways, the scope outlined in the White Paper was remarkably comprehensive as it placed administrators, faculty and students within one collaborative aggregation. The White Paper detailed four specific issues with campus-wide collaboration: incompatibility between some email clients; various scheduling packages used; sharing documents; and messaging. In bullet point, the document also listed six critical success factors and four technical considerations:

Critical success factors:

- Anytime
- Anywhere
- Email, calendaring, real-time collaboration
- One unified architecture
- Mobile capabilities – secure messaging anytime, anywhere

- Support single purpose clients like Eudora, Outlook Express, Netscape mail, Pine or any other standards-based mail client

Technical considerations

- Flexibility to deploy individual mailboxes, eliminating file contention and single points of failure. Support single mailbox backup and recovery
- Support Extended SMTP (ESMTP) for delivery notifications for Internet messages – reporting whether deliveries are successful, failed or are delayed
- Manageability – manage multiple servers from one unified administration tool along with Web-based administration. Ability to add users or groups and monitor mail from anywhere via a Web browser.
- Security – must offer an integrated public key infrastructure (PKI)

These key requirements were also prominent in a public presentation the office created. Where the presentation and the White Paper differed is in the inclusion of a brief presentation slide about competing products considered before choosing WorkWare. In addition to the specifications outlined above, the presentation also argued that OIT considered products used by other comparable academic institutions and by '[Northeastern's] top recruiters', acknowledging a certain amount of ritual corporate permeation into the decision-making process. The presentation concluded by forecasting a future Internet 'portal' through which campus users would access Web-based instructional technologies, groupware systems, Web applications and other administrative software (course scheduling, grades, advising).[3]

In an odd rhetorical choice, both documents situated OIT (specifically, Abigail) as the final arbitrator of the decision to choose WorkWare as the groupware product for the campus. The White Paper was overt about this decision-making role: 'After the analysis and review of many products on the market, I have found the best-fit product to meet both the critical success factors and the technical considerations for [the] University. The proposed solution for groupware on campus is WorkWare.' Unlike the White Paper, the presentation does not assign agency to the WorkWare decision and offers no insight into the decision-making process that resulted in WorkWare as the preferred option.

The agency invoked here strangely contradicted Martin's strategy to deflect his office's role as the final arbitrator of the groupware solution. At the same time, the actual institutional accountability for the WorkWare decision is more complicated. Several of my campus informants claimed that OIT was not the final agent in this decision but that the university president's office requested that OIT provide a rhetorical, retrospective technical accounting for what was the result of a corporate partnership assembled by his office. Such retrospective continuity, if true, belies much of the collaborative spirit with which the implementation was designed and announced and places Abigail's statement in a much different light. Regardless of the original architect, the sort of corporate-collegiate institutional aggregation ritualized here is nothing new to twenty-first century American higher education (White 2000; Faber 2003). Universities are partnering with corporations to sponsor athletic programmes and market the university as a sports franchise (Benjamin 2000), negotiate procurement deals and sell research technology (Kniffin 2000), contract out grounds, facilities and dining services (Dolgon 2000), construct course material and degree programmes (Jarvis 2001) and, as the WorkWare deal would have it, install and manage site-specific information technology environments for both administrative and pedagogical applications (Hanna 2000; Olsen 2003). In retrospect, Martin's key issue when he arrived on campus was in many ways the assemblage and dissemination of a problem for which a ready solution seemed to be already waiting.

According to the partnership outlined in the university president's email, CP agreed to donate software and two new servers to the university and, in return, the university agreed to adopt the company's new management information software, pay regular maintenance fees and purchase upgrades. WorkWare was designated to replace every faculty, staff and student's desktop communications and collaboration software. This meant that people on campus were expected to give up the software they were currently using to read, write and store email, schedule meetings, send instant messages, collaborate with multiple authors, and manage documents. In six short months, the institution was going to use WorkWare.

Social Relations: Organizing Time and Space

From: Announcements
To: Campus Community
Date: Fri 23 Aug 2002 13:55:33 -0500
Subject: WorkWare Implementation Schedule

[Northeastern Tech] is moving forward with the implementation of WorkWare. Over the next several weeks we will begin moving users from the current email and onto the WorkWare mail platform. We are following a schedule that will allow faculty and staff to be moved to the new system by 30 September 2002:

August 28: ROTC
August 30–September 2: Downtown Campus
September 3–5: Science Center
September 6: Library, Gym, Computer store
September 9: Campus Center Building
September 10–11: Admissions Offices, Physical Plant
September 12: Engineering Labs
September 13: Development Offices
September 16: Student Administration Services
September 17–19: Engineering Faculty Offices
September 20: Residence Life

Figure 3.2 WorkWare implementation email 2

The *domestication of time* is a central concept initiating and framing Simpson's critical study of technology and 'the price exacted by modernity in its scientific and technological guises' (Simpson 1995, p. ix). Such domestication appears in Figure 3.2, which presents a confident and succinct schedule for the WorkWare implementation. But, what does it mean for Simpson to argue that technology 'effects a domestication of time' (Simpson 1995, p. 4)? What is the relationship between technology and time and how might discourse innervate this relationship?

Simpson argued that time's domestication is prefigured by an emphasis on 'efficiency and control' (Simpson 1995, p. 4). In an effort to reduce, minimize and avoid time (repetition), efficiency and control break time into 'manipulable' and 'dispensable' units (Simpson 1995, p. 4). Simpson defined *domestication* as 'the will to control' and he wrote that time's domestication is fuelled by technology's 'rancor against the uncontrolled past and its concern with predicting and controlling the future' (Simpson 1995, p. 53). Describing two distinctive characteristics of the relationship of time to the technological project, Simpson argued that 'technology presupposes an awareness of the future', while at the same time 'promises a mastery that would "domesticate" the future' (Simpson 1995, p. 51). Technology, enfolded within the concept of progress and evolution, becomes a proxy for the future. As a proxy, technology reduces the amount of time needed for a specific task, enabling tasks to be accomplished with minimal time, with no time, without repeating time. This means that technology is disinterested about process on the way to its goal, or what Simpson called its 'means-ends' (Simpson 1995, p. 51). He accorded to technology an alternating ambivalence and animosity about process, since the end product of technology is 'something other than the activity itself' (Simpson 1995, p. 51).

This ambivalence leads to an externalizing and decontextualizing of process, 'just as the time of external history can be contracted without loss, the means can be decontextualized, taken out of time and world without loss' (Simpson 1995, p. 51). In other words, when time is shortened, reduced or even eliminated no substantial loss is inflicted to a system – in fact, the reduction of time is equated as a system-wide gain. As a result, 'action that is inherently meaningful in the sense that its very doing is meaningful' is taken out of the system prejudicially removed for the elimination of time (Simpson 1995, p. 51).

Simpson argued that when time is eliminated the future becomes 'condensed', controllable and, most indicatively, closed. By condensing the future, technology 'grant[s] us immunity from the uncertainty of time: the future would be predictable and controllable. In this way, that experience of time which we might call the "linear experience of time" – where the future is experienced as open, as the locus of possibility and the unexpected – would be "annihilated", in that the future would no longer be open' (Simpson 1995, p. 52). Otherwise put, when time is controllable, so too is human experience. The potential for randomness, chance and diversion is eliminated. When social relations become technologically mediated, communication, participation, consensus-making and other conditions for community are elided in the pursuit of ends.

Embedded in the shortening of process to the achievement of ends is an assumption that technology and its proponents share a suspicion and a deep anxiety about the future and human ability to control and predict future events. Technology becomes a supplement for human evolution, a way for humans to guard against or, at the very least, to effect some control over our own evolution and the anticipated changes that will occur in the world around us. In this sense, our ability to control time and space is also related to our ability to manage and control evolving social relations and networks around us.

Yet, as Latour (1996) reminded us, even if our technology is designed to 'safeguard' our evolution and our future, this future remains very much indeterminate (Latour 1996, p. 30). Once a discourse enables us to conceptualize and detach something different from the past and the present (a future), we then add to this discourse forms that enable us to speak in ways that influence or assert a degree of control over temporal difference. We may begin to approach temporality with confidence and certainty but this certainty only conciliates the 'gulf' between what Latour called 'the world of signs and the world of things' (Latour 1996, p. 24). Technology may be in the business of becoming, of collapsing signs and things, of creating things where nothing previously existed; yet, in technology's apparent temporal reduction, control and limits are still subject to real-time material events and conditions.

Here, we have the realization of what cultural critics call 'the simulacrum', by which we mean a constructed, simulated world (Baudrillard 1983). For some, the simulacrum is reason for concern and resistance, a call for a renewed authenticity. For others, the awareness of the simulacrum is unproblematically empirical, a descriptive accounting of how we emerge within and engage with our world. In each event, our understanding and participation within the simulacrum has important consequences for how we understand and interpret the social life-world in which we act. With the concept of simulacrum, we return again to the world of discourse and the constitution of the social by the discursive. The simulacrum is based on the idea of a linguistic code, but a code that does not refer to any specific referent. It is an image and a simulation, as Simpson put it, 'its point is not to make truth claims about the world, but is rather to construct for us a world' (Simpson 1995, p. 138). Discursively, the simulacrum is ultimately paradoxical. On the one hand, the simulacrum is problematic because it actualizes a language that is not referential. On the other, it is problematic because it assumes the possibility of a referential language; we are back to the desire for a language that effaces itself as a language.

As such, it is difficult to know what to do with the claim that 'there is no real world behind the symbols' (Krug 2005, p. 20). Although he laments the imposition of a supposed artificial virtual over a supposed authentic, Krug's account can be predictably ambiguous. He wrote, 'Within the technological system, people exist in spaces that are already constructed, and they have adapted themselves to this world as simply the way things are. Even nature is manufactured in the forms of parks and green spaces, at best engineered by civil planners to provide a simulation of nature in technology's conception of it: the optimal number of the most desirable trees, the proper amount and kind of grass etc.' (Krug 2005, p. 16). Although Krug desired an authentic instead of the simulacrum, more problematical for him is the ready acceptance and willing embrace of the simulacrum as somehow more natural than the authentic, the acceptance of 'a world picture ... in place of the world' (Krug 2005, p. 216). And yet, the point remains paradoxical. The world cannot not include technology, a problem that makes it difficult to parse the simulacrum from the authentic, if the authentic ever existed.

A partial solution to this problem is presented by Johnson-Eilola, who suggested that rather than constructing a necessary duality between authentic and simulated, we are situated within what Foucault called 'the epoch of simultaneity' (Johnson-Eilola 2005, p. 101; Foucault 1986, p. 23). Johnson-Eilola wrote that the epoch of simultaneity is an acceptance of juxtaposition, 'of near and the far, of the side-by-side, of the dispersed' (2001, p. 101). It is the acceptance of the authentic as simultaneously simulated, or at the very least already mediated. Johnson-Eilola wrote that this is a time of oscillation where time and space are both consumed and occupied (Johnson-Eilola 2005, p. 14). Otherwise put, it is where we claim to be suspicious of meta-narratives while at the same time willing to appoint technology as a meta-narrative to reoccupy their deposed space (Simpson 1995, p. 149).

Instead of disputing the authenticity of experience, what counts in the deployment and acceptance of the simulacrum becomes contests of value and power. As Simpson wrote, 'the discourses and practices that produce simulacra, like discourses of "truth" for Nietzsche and Foucault, and like "values" in general for Nietzsche, serve power' (Simpson 1995, p. 140). By making the simulacrum real, in Simpson's words, by 'procuring' the simulacrum, technology enables social control. The resulting challenge is not so much to reassert a natural or an authentic in the face of the simulated – since such authenticity is always already mediated. Instead, our discourses become directed to accepting or revealing the networks of power that have reconstituted time and space in their own self-interest. If, as Simpson wrote, 'the simulacrum is the postmodern form of value' (Simpson 1995, p. 139), we need to better understand what interests are served by the promotion of that value.

Social relations and time: the WorkWare implementation schedule

The most unique feature about OIT's self-imposed and now public mandate was that although OIT was tasked to install WorkWare campus-wide, and privately the university's president expected 100 per cent compliance across all campus groups, campus users were never specifically required to adopt or use the software. Given the institutional aggregation informing and directing the project, the plan's apparent concession to intellectual and academic freedom was surprising and perhaps even distinctive. As a consequence, OIT was accountable for the timely implementation of the first product from the university's high-profile corporate relationship but as agents of this transition, OIT had no real authority to force campus users to change software. Over the six-month implementation period that followed, this tension resulted in a rather dynamic rhetorical context. What could otherwise be termed a rather volatile simulacrum initially appeared stable and controlled. Yet, as our timeline unfolds, the simulacrum's initial collaborative instantiation will turn stochastic, then chaotic and then finally deteriorate into an all-out ultimatum (power).

In the time between emails 1 and 2, Martin and his staff remained confident in the implementation process. The White Paper created by his office, which outlined the advisory group's decision-making process and the eventual choice to adopt WorkWare, was provided to upper level executives at CP who promptly redistributed it to other college campuses where similar groups were reviewing WorkWare and competing software products. CP designated a 'critical care team' whose full-time responsibility was to support and enable the campus transition to WorkWare. Martin successfully directed two campus-wide open forums where he presented the White Paper, the three competing options and the documented reasons why WorkWare was chosen. Approximately 70 people from across the campus participated in these meetings, representing about one-third of the audience Martin needed to persuade to adopt WorkWare. Although I did not attend these sessions, Martin later recalled what he remembered to be a positive, unremarkable reception. Most participants were concerned about appropriate training.

At the time, Martin was most concerned about roughly 20 per cent of campus users who were dedicated to 'Pine', a Unix-based email platform, and 'Outlook', Microsoft's email program. Both systems were relatively stable, easy to use and well entrenched. Among the two groups, Pine users represented the more difficult problem because Martin's team decided the software for running Pine would not be available on the new servers. These users would be required to switch from a simple, stable, text-based, command line system to a new graphics-intensive, Web-based interface. To cast this disruption in protestant terms might appear contrived or perhaps facile. Though for many on campus such a depiction would not be entirely inaccurate as the change invoked a new semiotic and aesthetic for users who, for nearly a decade, had held out against graphic interfaces and the

Windows revolution. But, despite his anticipation, these users were not Martin's greatest problem as the implementation carried into the new academic term. As it turned out, the campus received a beta version of WorkWare. Therefore, the product that was buttressing OIT's new campus image and Martin's leadership credibility, the product campus users were about to be required to install, was not actually ready for implementation. This was not quite vapourware CP shipped Martin an actual product. However, it was delivered more or less as a work in progress. As people started to use WorkWare, they quickly discovered that the software did not work. To be less than subtle, this materiality seriously disrupted Martin's projected timeline and his installation deadlines.

Martin and his staff claimed that they realized that WorkWare was beta only a week before the implementation was scheduled to start.[4] His exasperation was still evident some time later, when he told me that his office 'had gone through all this work', they were determined to see WorkWare run. Martin's commitment to his simulacrum was acute and all-encompassing. The project had campus buy-in, a high-profile corporate partner in CP, a presidential directive to install the product, a popular White Paper making the rounds at campuses across the country, and an on-campus customer-service image still in fabrication. As they approached implementation, Martin and his staff needed WorkWare to work. And, it is important to realize and admit that from his perspective, to this point in the implementation, the technology did work. The term 'moving forward' inscribed in email 2 was as much a statement of confidence from Martin to his staff as it was from OIT to the university campus. Martin hoped that the email could restore some confidence among the implementation team as he and his staff saw few options beyond simply willing WorkWare to operate successfully.

Beliefs, Values, Desires

From: Announcements
To: university_community
Date: Fri 20 Sept 2002 15:51:05 -0400
Subject: WorkWare Update

The first phase of the WorkWare migration is approaching completion. All buildings have been visited, but not all requesting faculty and staff have been migrated. The last week of September and the first week in October have been reserved for the migration of everyone who remains. To help us in this effort, please contact Kendra Smith <ksmith@university.edu>, if you have not done so already, to confirm that you wish to be migrated now.

Northeastern Tech will have a profound impact on the final WorkWare product. Your comments and questions are important to CP. The following represents a portion of the problems you are sharing with us.

1.WORKWARE CRASHES. The server stops working at least once a day during the week. Outages last 15–20 minutes. CP has identified and recreated this problem at their facility. This has been escalated to a high priority for resolution at CP.

2.TEXT IN RECEIVED MAIL RUNS OFF TO THE RIGHT IN A LONG STRING. This appears to be a problem with Internet Explorer version 6.0. CP is aware of this problem and is working to correct it.

3.TEXT INCLUDES HTML HEADERS. Email sent in rich text (which is the WorkWare default option) sometimes displays all the formatting information, which renders the message illegible. This is not a consistent problem. Plain text seems to be OK, but you lose special text settings like colour and bullets, etc. This is also a high priority and CP is researching this concern.

Additional comments and answers are available at <www.university.edu/~workware/issues.html>

Many of you have communicated enhancement requests to our office. A portion of this list appears below. We are forwarding this list to CP.
– Allow users to add words to the dictionary
– Allow users to spell check all messages without having to select the feature each time.
– Allow users to save multiple signature files and pick from a list if default isn't desired.
– Allow users to move from one message to the next or previous without returning to the listing of messages in the folder.
– Allow users to hide the membership of user-owned contact groups.
– Anti-Spam button that would automatically configure a rule to delete any email with the same address of the current message
– Audible alarm

Please do not hesitate to share your suggestions and frustrations; call Kendra at x4321 or write to us at <workware@university.edu>.

Figure 3.3 WorkWare implementation email 3

By late September, when email 3 was written and sent to the campus, OIT was losing the support it had aggregated in July and August as exponentially more users reported technical problems with WorkWare. 'There was quite a bit of frustration' Martin recalled to me with considerable restraint. He noted that this frustration was evident in both the campus community and within OIT as Martin and his staff realized that their carefully reconstituted identity was now entangled with a new aggregation: a supposedly faulty product, lost email, system outages and unreadable messages. Even the relatively basic calendaring function lost appointments and would not allow assistants to place meetings on their supervisors' accounts. Martin's campus support was also eroding. Members of his campus advisory committee stopped coming to meetings, CP's critical care team had dissipated and university officials claimed that WorkWare was Martin's personal conceit.

Martin's overt Star Trek metaphor seemed suitably heroic and team building for a group of computer programmers in distress: 'our engines were down' he recalled. More to the point, he reluctantly acknowledged, 'CP let us down and we were hung out to dry'. At this time in the implementation, Martin was still reticent to believe that WorkWare did not work. Like any software, the technology had bugs but he insisted to me that no email was ever lost. His staff at times did need to recover emails that were misplaced by the new servers but he was unyielding in his insistence that no email was ever actually lost.

Of course, technology and the decision to use a specific technology is never value-free. But it is important to reconsider the perhaps better-recognized story which claims that technology independently emerges from and promotes its own values which are typically claimed to be things like efficiency, rationality and control. Technology is a means employed toward the achievement of a specific, predetermined end. Thus, it is important to ask to what ends are efficiency, rationality and control deployed?

Technology does not, on its own, introduce unique values or beliefs within a social system. Instead, it represents a choice to allow specific values and beliefs to become actualized as end-products. At the same time, a choice to use technology discounts other values that might inform how a process will take place or how a goal may be achieved. And, more problematically, the uses of certain technologies represent choices to devalue what could be meaningful experiences. Rather than choose a process that might endure over time, that might initiate tangential experience, that might include learning outcomes, or influence social relations, the choice to employ technology erases these possibilities in pursuit of an ultimate goal.

When Krug correctly argued that 'the order that may be found in this world [technology] exists not for the moral improvement of the human but for the further extension of the system itself' (Krug 2005, p. 14), he asserted that using technology invokes a choice that often results in further technologization. This choice reflects what Simpson called a 'value' by which he referred to a 'means-ends' choice (Simpson 1995, p. 5). For Simpson, 'value' is something distinct from 'meaning'. Whereas value represents a privileged choice, 'an arbitrary product of the will' (Simpson 1995, p. 4) or an anticipated goal we hope to achieve (Simpson 1995, p. 43), meaning resides outside of value choices as something that cannot be engineered, as 'practices whose significance cannot be duplicated by a functional equivalent' (Simpson 1995, p. 165). 'Meaning', wrote Simpson, is something 'found not made' (Simpson 1995, p. 165). Simpson's distinction between value and meaning roughly corresponds to an idealized and perhaps naive distinction between technology and science when science is characterized as primarily explanatory and technology as interested in using explanatory propositions as rules for action. Meanings encompass empirical experience of phenomena while values choose or dictate the end result of how phenomena will be used.

Understanding the meaning/value distinction is important to understanding technology because technology often causes us to confuse meaning with value (Simpson 1995, p. 48). Because any given technology is focused on achieving an end-product, it becomes too easy to confuse the way in which we achieve that end-product as meaningful. In choosing rationality, efficiency or utility as a way to achieve a goal, we confuse these values with meanings. Efficiency is not meaningful, it is a procedural choice based on a predetermined value. Simpson noted that 'technology is able to provide for us a state of satisfaction' but as we become more proficient in securing the end state, we forget what we found meaningful in that end state and we do not see the process of reaching that end state as significant. The process 'has been demoted to the status of merely means' and the end state becomes an alienated satisfaction – an end state that is broken off from the reasons for its realization (Simpson 1995, p. 164). Technology's values are incompatible with meaningful life–world experiences because technology will always seek to eliminate experience, process and meaning on the way to achieving a goal.

This incompatibility with meaningful life–world experiences bases many cultures' deliberate rejection of certain technologies. For example, over the past decade, a community of conservative Amish have moved to communities several miles from Northeastern Tech. Though famous for their rejection of modern technology, that claim is somewhat exaggerated. These Amish use technologies, their buggies, farm implements, wood stoves, all representing certain (if dated) technologies. However, they consciously reject technologies that would enable them to live independently from each other (destroy dependencies that create community), would initiate competitions for status among each other or lead to vanity. Here, technology is not sinful, but as Howard Rheingold (1999) wrote in *Wired Magazine*, the decision to value and use a specific technology is mediated by the influence the technology will have on what one believes is meaningful and valuable. When asked about using a mobile phone – but only from a booth in a meadow – Rheingold's Amish host stated, 'We don't want to be the kind of people who will interrupt a conversation at home to answer the telephone. It's not just how you use the technology that concerns us. We're also concerned about what kind of person you become when you use it' (Rheingold 1999, p. 2). Rheingold concluded that 'far from being knee-jerk technophobes' the Amish he interviewed 'devise remarkable technologies that fit within their self-imposed limits' (Rheingold 1999, p. 2). Such self-imposed limits would appear to ensure that technology does not interfere with meaningful life experience nor that the use of technology would be inspired by values the larger community rejects.

When meanings are supplanted by values, it becomes difficult to understand or remember the reasons for technological choices of utility, economy, rationality or efficiency. Notice that the reduction of meaning achieved by eliminating process or life experience from activity correlates with the reduction and the domestication of time as discussed above. All technological ends originated in human choice and human desire, even though these desires are often forgotten in overwhelming technological systems and in the massive cultural and social apparatus (institutions) that can be built to sustain any technological system (Simpson 1995, p. 13). 'To look at the world technologically', wrote Simpson, 'is to see it either as resources to be used or as constraints to be overcome' (Simpson 1995, p. 15). This gaze has been preconditioned by an already-articulated end and must therefore operate in service to that specified end. In this way, a technological project is 'autonomous with respect to traditional values' (Simpson 1995, p. 19) because the project is answerable only to technical values of procuring the desired end. Other values, 'estimates of worth' and meaningful experiences are irrelevant to the predetermined task at hand. Because the end-goal has already been predestined, technology has full licence to bring the stated goal into being at any cost to what may be valued, believed or held to be meaningful. But, it is important to realize that this is not technocratic determinism. Humans establish (and then often forget) the values that the technology was designed to fulfil.

Values, beliefs and meanings: 'Northeastern Tech will have a profound impact on the final WorkWare product'

A surprise user-test was announced in email 3, paragraph 2 of the WorkWare implementation announcements. As the implementation began to fail, Martin approached the campus with a new narrative. He wrote, 'Northeastern Tech will have a profound impact on the final WorkWare product. Your comments and questions are important to CP. The following represents a portion of the problems you are sharing with us.' As noted above, that sentence was followed by a listing of some of the various difficulties campus users were experiencing with WorkWare. Following the list of 'bugs', Martin wrote, 'Many of you have communicated enhancement requests to our office. A portion of this list appears below. We are forwarding this list to CP.'

The user-test was a strategic move designed, ironically, to buttress the now faltering WorkWare campaign by further extolling its connection with the newly absent corporate partner. According to this unfolding narrative, Martin's staff revealed that the campus received such a good deal on WorkWare explicitly because they were user-testing the new product for CP. In return for their feedback, the campus would know that they helped to solidify the next iteration of CP's groupware suite.

In retrospect, Martin admitted that the user-testing strategy was a fiction rapidly designed at the time to stabilize the ground crumbling beneath him. However, and he is adamant about this, he claims he did send CP a list of campus issues and requests, even if this feedback was not planned within the original blueprint for the WorkWare transition.

Discourse

In the preface to his book *Tracing Genres Through Organizations*, Spinuzzi (2003) described writing as the product of necessary, task-based innovation: a way to alleviate local problems. Citing Denise Schmandt-Bessarat's (1986) study of Sumerian writing, Spinuzzi wrote that 'perhaps the most powerful and protean tool we have devised, the one so many ancients believed to be a gift from the gods – can be traced back to a series of slapdash innovations largely meant to ease the lives of clerical workers' (Spinuzzi 2003, p. ix). In other words, writing emerged as a technology. Of course, not all discourses take the form of written text. Yet, most are typically motivated to achieve certain goals, as Simpson noted, 'words have not only meaning but also function' (Spinuzzi 2003, p. 125). As such, we can, to a certain extent, consider discourse as a type of technology. At the same time, discourse represents a temporal process, a way of engaging with the world, with social and material structures, and a way to create value and meaningful experience. As such, we can consider discourse part of the simultaneity that for Johnson-Eilola defines the postmodern era. Certainly, accounts of discourse and technology have depicted discourse as both fully determined by technology and fully resistant of technology's solipsisms.

Krug took a typically deterministic perspective toward technology and discourse, arguing that 'the language that frames our experience of the world and with which we write our histories has changed dramatically over the last few centuries' (Krug 2005, p. 11). This language, according to Krug, no longer emerges from local human experiences, 'elaborated through conversations with other people' (Krug 2005, p. 11). Instead, it comes to us 'pre-packaged', reflecting not local human experience but 'the values of capital, the machine, and the technological system' (Krug 2005, p. 11). Citing Denzin (1991), he argued that technology now tells us who we are; we know ourselves from the pictures we gain from technologies like television, the periscope, microscope, CAT scan and the camera. This happens because technology makes local language practices 'inadequate' for understanding and communicating experience. We allow medical imaging to define us through its ability to visualize inside the body (Krug 2005, p. 21). Soon, all forms of (self) expression are subordinated to technology: 'After Warhol the magazine becomes a parody of its own immolation in an exclusive concentration on style, technique, and the immediate. House, trance and techno music reiterate the rhythms of the machine, while the contrary examples of punk, grunge, and so on are, in their beginnings at least, reactions to the constraints of the system and, after that, simply more, commodified, prepackaged forms of expression' (Krug 2005, p. 21). After technology, there is no possibility for discourse to escape the system.

For others though, Krug's entrapment can be viewed quite differently. To personify, technology may want to subdue discourse but it remains constituted by and dependent on discourse for its power. Thus, technology is caught in the dynamic of being constituted by and yet constituting discourse. Johnson-Eilola elicited this perspective when he cautioned against claiming that a specific technology 'somehow autonomously causes broad social changes' (Johnson-Eilola 2005, p. 34). For him, technology does not drive social change so much as it works in concert (simultaneously) with multiple social, material, and incidental forces. For examples Johnson-Eilola cited Gibson's (1984) cyberpunk novel *Neuromancer* which was highly influential among programmers and interface designers. Yet, Gibson himself wrote the book on a typewriter and had little experience with computer technology (Johnson-Eilola 2005, p. 34). Similarly, Johnson-Eilola listed advertisements, film and other forms of media as discourses that 'exert strong tendential forces on the development of computer interfaces' (Johnson-Eilola 2005, p. 34). Thus, technologies 'participate' in social systems in what Johnson-Eilola calls 'recursive and contingent processes' (Johnson-Eilola 2005, p. 34). Technologies rearticulate social, material and institutional structures and processes and this rearticulation is shaped and 'intimately bound up' with discourse and discourse practices (Johnson-Eilola 2005, p. 17).

As a consequence, Johnson-Eilola argued that the 'noise' often made about technologies and technological determinism 'frequently blinds us to more profound shifts in our culture – changes that may have, in the long run, a larger effect on our futures as well as our current lives' (Johnson-Eilola 2005, p. 8). These changes may employ technologies as a means-goal, or as will be demonstrated in the case studies in Chapters 4 and 5, they may employ discourse as a specific technology. The point here is that in our concern about oppressive technologies, we often lose sight of the new discourses and social formations the technologies inscribe and enable.

Email 4 (next page) was sent directly to individual campus computer users who, midway through the implementation process, had not been visited by OIT staff. Along with email 1, which was addressed personally by Martin, it is the only other email addressed from a specific person. However, that person had now changed from Martin to his subordinate Kendra, the new WorkWare point-person. Kendra Smith, the author of email 4, began working for OIT mid-summer and did not have a role in the consensus-building exercises that prefigured the WorkWare implementation. She first appeared during the implementation as the contact person listed in email 3 and is now the third WorkWare point-person in the four-month implementation period.

Email 4 was sent after a full 49 days, nearly a month and a half, of silence. Gone from the implementation campaign is any mention of Martin's sudden scheme to turn campus problems into 'a profound impact' for WorkWare's final instantiation. There are no references to CP, the corporate partnership, or to the university administration's grand plans. But, added to the campaign message is the significant concession that users do

Date: Fri 8 Nov 2002 09:58:19 -0500
From: Ksmith@university.edu
To: user@university.edu
Subject: Email

The Office of Information Technology requests that you contact us with your email requirements.

If you would like to be migrated to the new WorkWare system, if you would like WorkWare training, or if you would like to discuss your problems or concerns with WorkWare, please contact Kendra Smith x4321 or Ksmith@university.edu.

If however, you plan to remain with your current email client we still need to move your mail service to the new servers WWS1 WWS2, the servers used by WorkWare. This is necessary because the current mail server in use by Northeastern is over four years old and is starting to show its age, especially with the increase in email traffic over the past several years. The process simply changes the location of where your mail is picked up and delivered. This will require a technician to visit your desktop but it is a painless process, and in most cases, will be totally transparent. Contact Kendra to set up your appointment.

If you have specific questions pertaining to this switch, please call the helpdesk x1234 and we will have a technician work with you to explain the changes. In order to better serve the campus community we request that you share the reasons you have for not moving to the new system. Please forward your comments to Kendra.

Kendra Smith
Communications Training Coordinator
Office of Information Technology
Box 6789
University
(789) 123-4321

Figure 3.4 WorkWare implementation email 4

not have to use WorkWare. The concession is remarkable in its conventionality. The WorkWare campaign was built on the deliberate suppression of alternative choices. Early planning documents (Feb 2006) noted that although WorkWare was intended to be 'the de facto supported e-mail system' at the university, WorkWare would be compatible with third-party email clients. Significantly, the planning document argues that OIT should choose not to support other email clients. This proposition echoes the initial rationale for WorkWare which was to consolidate the array of dissimilar communication systems used at the university:

> The WorkWare environment (including email, described below) is intended to be the de facto supported email system at [Northeastern] University. This extends to all computing users both on and off campus. The WorkWare

> implementation team however realizes that some clients may wish to use their own email clients. The WorkWare environment provides POP3 access to users' electronic mailboxes, and is therefore compatible with the vast majority of third-party email clients.
>
> Users of third-party email clients should realize, however, that specific value-added services which are part of the WorkWare system will not be available to them. These include items such as global address lists, and integration of email with calendaring and scheduling features. Support levels for users of third-party email clients are equally limited, as Campus Information Services will provide total problem resolution only to WorkWare email users, with problem resolution of other email clients on a best-effort basis. (implementation document, 6 Feb)

As one member of the implementation team discreetly told to me in a separate interview, early in the planning stages of the implementation campaign there was a 'conscious choice to silence other options'. The strategy was to get people on board and then offer alternatives.

Delivered in retrospect, my informant's claim suggested a double strategic intent: a description of the implementation strategy overlaying a convenient after-the-fact explanation for the shift in policy. The actual concession of email 4 was slightly backhanded, 'if however, you plan to remain with your current email client'. But, given the nearly two years of preparation, this was a dramatic offering. What began as a 'unified architecture for email, Web access, online calendaring and group scheduling, and collaborative workspaces' (implementation documents), and an opportunity to realize 'unprecedented concepts that have never been fully realized or implemented at [the] University' (implementation documents), had become a new server implementation.

Email 4 was designed with the singular purpose of moving campus email to the new servers donated by CP. Martin alleged this to be 'a friendly reminder to get off the hardware'. He would have liked everyone to be using WorkWare, but he realized that this was no longer a realistic ambition. At the writing of email 4 (8 November), Martin's staff were concerned that a number of campus users were not communicating with OIT or were not paying attention to the ongoing WorkWare campaign. He estimated that email 4 was individually directed to about 20 per cent of the campus, mostly members of the university's faculty. OIT staff used the term 'hold-outs' when describing these users, noting that their resistance was more intentional than ignorant. Yet, at the end of the semester (18 December), OIT was planning to switch all email service to the new servers. Therefore, even if these 'hold-outs' refused to use the WorkWare software, their email software (client) needed to be directing their email to the new servers.

As noted above, the concession to third-party email clients was significant. The implementation planning documents acknowledged that the WorkWare hardware infrastructure was compatible with third-party systems. This was posed internally as a strength of the WorkWare platform;

but until now it was not mentioned in the public communications. Email 5 (reproduced below) was sent to the university campus approximately five hours after email 4 was sent to individual 'hold-outs.' It was the public version of the individually addressed email 4. Martin noted that email 5 was created when his staff questioned if their list of 'hold-outs' may be incomplete. They noted that some faculty may have been off campus, or not paying close attention to university email. His staff also reasoned that the campus could use reminders about WorkWare training.

From: Announcements@university.edu
To: university_community
Date: Fri 8 Nov 2002 15:09:54 -0400
Subject: WorkWare Update

All faculty and staff members who have expressed interest in moving to WorkWare have been migrated. If you have not been migrated and would like to be, please call x1234 or Kendra Smith (x4321). We will be happy to schedule a time that is most convenient for you.

Training (and/or retraining) is available on an individual basis or with small groups within departments. This training occurs right at your desk. It can be very basic or it can be a time for you to discuss your specific needs and problems concerning WorkWare. We can also set up some of the preferences that you may have overlooked. These sessions have been very positive and productive. Many of the difficulties users face can be addressed during these sessions. Other, more difficult problems will be handled within a few days. Call the helpdesk or Kendra to schedule your appointment. Please do not hesitate to communicate your needs to us.

Currently, over 400 faculty and staff members have migrated to the new WorkWare application. Our next step will be to work with those who have not converted. We will contact you to discuss the concerns that you have. Once we have all these reasons documented, we will try to find solutions for each issue.

Additionally, we will be migrating non-WorkWare users to the new server. The current mail server in use by the university is over four years old (that's 80 in computer years) and is starting to show its age – especially with the increase in email traffic over the last several years. We need to move all mail servers to the new servers WWS1 and WWS2, the servers used for WorkWare. This process simply changes the location of where your mail is picked up and delivered. Therefore, users who choose this option will still be able to use their current client. This process is painless, and in most cases, will be totally transparent.

If you have specific questions, please call the helpdesk, and we will have a technician work with you to explain the changes.

Figure 3.5 WorkWare implementation email 5

'What counts, in a technological project,' wrote Latour, 'is deciding what has to be negotiated, and deciding on an official doctrine that will make it possible to proceed with any negotiation at all' (Latour 1996, p. 112). At their core, technological projects are not about machinery, technique, rationalization, efficiency or utility. Latour wrote, 'in the beginning there is no distinction between projects and objects. The two circulate from office to office in the form of paper, plans, departmental memos, speeches, scale models, and occasional synopses. Here we are in the realm of signs, language, texts. In the end, people, after they leave their offices, are the ones who circulate inside the object' (Latour 1996, p. 24). If technological projects have a core, it would be composed of persuasion, consensus-building, negotiation, storytelling, empathy and argument.

Technological projects are about discourse, or in Simpson's terms, about 'language games' (Simpson 1995, p. 103). Borrowing the term from Habermas, Simpson suggested that to fully participate and engage in social situations (like technology development) requires an understanding of what it means to engage in argumentative conduct. The strategic deployment of discourse thus requires 'that the observer draw upon some participative appreciation of that language' (Simpson 1995, p. 103). 'We would not know what it means to engage in argumentative conduct, and hence would not be able to identify such conduct, without our already having acquired the experience of what it is like to participate in argumentation' (Simpson 1995, p. 103). Thus, to successfully participate in a technological project requires that a person have acquired the experience of discourses that construct, restrain, promote and aggregate that project. It requires that participants 'draw upon some participative appreciation' of the specific language games that constitute the project (Simpson 1995, p. 103).

And yet, to return to Latour (1996), what is most often excluded from and viewed antagonistically to the study of technology and technological projects is language. Latour wrote, 'people who study technological projects take too little interest in the official doctrines dealing with the actual management of projects. This metalanguage appears parasitical. Yet it plays the same essential role that strategic doctrines play in the conduct of wars. In the course of a battle or a project, ideas about the way to handle battles or innovations play a performative role' (Latour 1996, p. 113). As much as technology would hope to subdue discourse, to turn it into a simple conduit for information, technology requires discourse to be performative, constitutive, enabling. Otherwise, technology has no ability to operate.

'Frustration' is the word Martin selected to summarize his mood as he and his staff composed emails 4 and 5. 'Some people wanted to call it off,' Martin sighed. Indeed, several prominent, vocal faculty were posturing in campus meetings that Martin, OIT, WorkWare and most of what Martin had crafted together over his first year and a half in the job, were incompetent. These were faculty who believed that their community allegiances and reputations started at their Internet portal. What happened on campus was of relatively little interest and Martin's little distraction had grown to the point that it was now threatening to detach these great physicists and engineers from their worlds of influence. More sympathetically, their most pressing problem over the past month was that WorkWare was losing emails, a problem made more interminable because the victim does not even know that intended email was never delivered. After some digging through electronic files, Martin claims, his staff were able to find where these errant messages were going. 'We did not lose control of email,' Martin insisted, 'but it was clear the user community thought we did.' Emails 4 and 5 were designed overtly and primarily to respond to this real and anticipated loss of support for WorkWare across campus.

Kendra was not designated as the author of the more institutional email 5 but she was listed as contact person. The messages of emails 4 and 5 were largely equivalent: 'Everyone we knew who wanted to be migrated had been migrated, let us know if you haven't'. Martin's confidence was unshakable. In retrospect, there were good reasons to call the implementation off. But Martin is not that kind of leader. 'We were getting calls from all sorts of people who didn't know the level of support we provided.' He realizes, now, with some distance, that the frustration permeated across all levels of the implementation. His position is unalterable though, support was available and if people would simply communicate with the implementation team, he would personally come to their desktop and sit with them through their problems. 'The users' difficulties can be addressed,' he insisted, unconsciously using present rather than past tense to describe his state of mind at the time.

But, Martin's confident self-assurances are more genuine in retrospect. The public version of the 8 November announcement invented a new category, 'non-WorkWare users'. Like the private concession inscribed in email 4, which allowed users to 'remain with your current email client', the new category signified a radical, public shift in policy. Unlike the morning's email, this message was sent to the entire campus community and as such required a certain degree of face-saving. Thus, the concession was not as generous in this version. Before electing to remain with one's own email client non-believers (those who have not converted) needed, at least in the public declaration, to confess their concerns with WorkWare. Once reasons for not converting have been documented, the email states, OIT staff will 'try to find solutions for each issue'. Only after this procedure does the email admit the new subject 'non-WorkWare users' who will be migrated to the new server. Further, email 5 explicitly states, 'therefore users who choose this option will still be able to use their current client'. Martin is unusually subdued about the decision to offer users a choice this late into the implementation. He reiterates that the policy remained to 'get as many people onto WorkWare as possible'.

Power and material practices

From: Announcements@university.edu
To: university_community
Date: Fri 6 Dec 2002 15:47:23 -0400
Subject: Migration to WorkWare Email Server

Over the past several months, the Office of Information Technology (OIT) has been migrating the email accounts of University email users over to the new WorkWare email servers. Approximately 70 users have not yet been migrated.

As of 18 December 2002, all campus mail will be redirected to the new servers. Regardless of the email client you are using, if you have not made arrangements to have your mail service cut over to the new system prior to this date, you will not receive any new mail after 18 December 2002.

To prevent this loss of service, please contact the helpdesk x1234 to make arrangements for your mail service. We do not want anyone to lose email service.

Figure 3.6 WorkWare implementation email 6

Email 6 was written as a penultimate warning to the 70 people who had not yet responded to the WorkWare campaign. Martin said that at this point, 12 days before the old servers were to be turned off, he wanted to make sure no one was left out, 'that no one loses their email'. At this point, the implementation campaign was essentially complete and the messages in emails 6 and 7 demonstrate the impending finality of the project. The scope of the implementation has also significantly retrenched. The WorkWare software is not even mentioned and the campaign is now about installing servers. The author, who is not mentioned, has adopted an agnostic concern about users' email preferences, noting that 'regardless of the email client you are using' computer users on campus need to make arrangements to use the new servers provided by CP.

This email is also the first to signify a termination date for the campaign – and for unconverted email accounts – 18 December. As will be further examined in subsequent chapters, agency for continuing one's email service was shifted onto users and OIT appeared to no longer take responsibility for the continuation of email service, the email software people decide to use on campus, or any trouble WorkWare may be causing. In addition, OIT appeared to have abandoned the grand statements, policy decisions and corporate partnership benefits of earlier communications. Far from the optimistic and glowing proclamations of email 1, email 6 foreshadowed the final perfunctory ultimatum of email 7, which terminated the WorkWare campaign:

From: Announcements
To: University_community
Date: 17 Dec 2002 10:50:49 -0400
Subject: Old E-mail Servers Go Off-line Tomorrow

If you have not responded to any of the requests OIT has made over the past five months regarding the changing to the new WorkWare email servers, you need to call them by 4:30 p.m. today. The old email servers will be turned off at 6 a.m. tomorrow, Wednesday 18 December. No new mail will be directed to these machines after this time. Please contact the helpdesk x1234 with any questions. Thank you for your cooperation during this transition.

Figure 3.7 WorkWare implementation email 7

Email 7 completed the WorkWare implementation with an ultimatum. As with the previous email, there is no mention of the WorkWare software. Instead, the email retroactively claimed that the preceding five-month campaign was designed to move campus computer users to new servers. Those who have not switched servers will not receive email when the old servers are terminated.

Thus, the WorkWare implementation concluded with a final act of materiality (physically turning off the old servers and turning on the new ones) and power. Absent at its conclusion was the sort of heterogeneous moments of instantiation and social creation Harvey's framework suggested construct a mediated context. In the absence of an institutional network, of the ability to manage or influence social relations, of the realization of value or the creation of meaningful experience we also see the absence of any particularly persuasive discourse. We are left with power and materiality, oddly enough in the form of a proclamation, the same genre that initiated the implementation four and a half months ago.

At the same time, although the WorkWare implementation was not fully successful, a different OIT emerged at the end of the campaign. When I met with him about these final two emails, Martin did not have much to say about these messages. I don't read his silence as failure or regret so much as exhaustion and a certain amount of disinterest in an ambition that he claimed was never properly supported or enabled by CP or the university. In the future, he claimed that the implementation was successful and in the next six months his office introduced and refined the anticipated information technology portal which moved faculty grading and student registration, course selection and housing assignments to an online interface into which access to WorkWare was embedded. WorkWare did not achieve the 100 per cent migration the President expected but soon he too moved on, retiring to a distinctly low-tech life of fly fishing and camping in a newly purchased recreational vehicle.

Chains of reason: social positioning and social change

'The work of generating interest consists of constructing these long chains of reasons that are irresistible, even though their logical form may be debatable,' wrote Latour (Latour 1996, p. 33). WorkWare did not secure such 'chains of reason', but along this implementation other chains were constructed that in retrospect appear far more durable and irresistible than a poorly constructed email system. In all the 'noise' that was created around WorkWare, few people paid attention to, or did not sufficiently notice, the renewed social position OIT attained on campus. Latour noted that people often wonder how technology has a social impact, 'how a laboratory, or a science, can have any impact at all on society' (Latour 1996, p. 33). How is it that something relatively obscure or cloistered can have any influence on daily life in an organization? a community? a population? Latour argued that the answer to these questions can be found in what he called 'chains of translation' which transform meta issues into micro ones, or in his words, 'global problems (the city, the century) into a local one (kinematics, continuous transportation)' (Latour 1996, p. 33). These methods of translation are not necessarily logical but they 'oblige those interested in the global problem to be interested through almost imperceptible shifts, in the local solution' (Latour 1996, p. 33).

At the end of the WorkWare campaign, we can argue that, at least in the global/local dynamics that occur within the microcosm of an organization, translation can work both ways. Translation can oblige those interested in a micro problem to be susceptible to a global solution. To be explicit, the campus leadership's interest in securing the CP partnership and its manifestation as the local implementation of WorkWare established the conditions whereby OIT could reconstruct itself as a powerful social agent on the college campus. Thus, at the end of the campaign, OIT is no longer implementing WorkWare, it is implementing itself as the sole arbitrator and manager of computer technology on campus. The message of the last two emails instantiate this social position. The 'heterogeneous moments of instantiation and social creation' that would have supported WorkWare have aggregated instead to support OIT's institutionalization on campus.

Simpson's theory of technology and meaning casts a subtle, ironic light on this twist in the WorkWare implementation. He argued that 'through its own domesticative practices, technology effects a shift in our attention away from practices that would make a meaningful difference' (Simpson 1995, p. 167). In this case, WorkWare technology effected an ironic shift in attention away from the more meaningful work OIT was doing to create a technological dependence on itself across campus. Martin emerged from the implementation context with two separate agendas, the de-legitimized WorkWare implementation, and the larger project to gain credibility and authority for his office, OIT. The implementation offered Martin the opportunity to gain a more authoritative and empowered social position on campus.

To achieve this position, Martin did not need WorkWare to succeed so much as he needed to convince users that they were dependent on OIT for their information technology needs. This overt claim was announced early in the implementation (email 1) and it was supported discursively through to the end of the final email (email 7). Martin created this dependency materially, but also discursively, by insisting that users required OIT to enable their computer technology. Recall that the emails never gave users the agency to switch their own email service to the new servers. Instead, they required that users depend on OIT for this relatively small fix. Here we can see actions that radically influenced this social culture that were never overtly stated or semantically communicated. Instead, through specific discursive activity, supported by material action, Martin was able to reconfigure his position on campus and create a network of power for his office.

The following chapters will elaborate this discursive activity and the various functional levels on which this discourse shaped the renewed social network on campus. For now, we can briefly review several themes asserting OIT's authority and campus legitimacy and power. Note these themes from email 5 (which will be further developed in Chapter 4):

> We can also set up some of the preferences that you may have overlooked.
> Our next step will be to work with those who have not converted.
> We will contact you to discuss the concerns that you have.

We can see in this use of thematic instantiation a grammatical repositioning of OIT that asserted a new and more essential campus position. This interweaving of discourse, materiality and power constructs a more meaningful social network on campus. The materiality of the network (the fact that new servers are being installed) means that both success and failure of WorkWare as a new software platform furthers the campus's dependency on OIT. In the competing agendas that constituted the WorkWare implementation context, the actual success of WorkWare suddenly counted for very little. Does this mean that Martin intentionally sabotaged his own technology? Did he deliberately allow WorkWare to fail? Probably not. In speaking with Martin, I have no reason to think that his attempt to install WorkWare, to work with CP, and to implement the President's plan were not genuine.

However, this does not mean that strategically, over the changing dynamics of the campaign, Martin did not deliberately or even unintentionally take advantage of an opportunity. As I argued earlier in this chapter, technology enables pre-existing beliefs and desires to become evident through a direct means–ends articulation. Simpson wrote that 'in understanding an expression or an action, we look for reasons – typically beliefs and desires in the context of socially sanctioned rules – that underlie it'. Rhetorically and strategically, we must admit more than logic or rationality as motivations for human action. As Simpson claimed, 'we must evaluate [beliefs and desires] as *reasons*, that is, as being *understandably* connected with the utterances or action in question' (Simpson 1995, p. 105, emphasis in original). Martin's actions may not have been deliberate, rational or particularly strategic, but as Latour noted, 'technological reality isn't rational, and it's no good rationalizing it after the fact' (1996, p. 37). Latour reminded us that we must not believe in the autonomy of technology. Technology is 'weak and fragile' (Latour 1996, p. 292), it is dependent on our discourses for its power and on our values for its motivation. We may use technology to elide meaningful experience, to erase time, to condense social space, or, like Martin, to build a powerful empire within our workplace. But, these are our choices, not those of an independent and autonomous technology.

This chapter presented a narrative reconstruction of the email campaign OIT used to implement WorkWare at Northeastern Tech. As should be evident by now, my interest in reconstructing this story is not to report on the ultimate success or failure of the campaign, nor to present advice on how such campaigns should be managed. Nor do I wish to judge or hold Martin and his staff accountable to various ethical or managerial postulates. Instead, I hope that in reviewing these documents the next chapters can initiate an exploration and problematization of the formal, functional and ethical characteristics of discourse during moments of change.

As I noted previously, the central purpose of this book is to initiate an examination of the discursive features of social change and to explore the specific kinds of discourse that enable social systems to change. Ideally, this endeavour will lead to a space in which we can critique and discuss the operations and problematics change from a position that enfolds an understanding of the discursivity of change within a critical examination of material actions people will take when tasked with implementing, directing, or managing change.

Symbolic activity is crucial to enacting change as the reconstitution of a disrupted and dysfunctional social system. To use the vocabulary introduced in the previous chapters, my principal questions as we move towards the more analytic sections of this study are broadly categorized under the larger question: What are the effects of certain discursive forms on social organizations and structures? As components of this larger question, the next chapters question: What features of discourse enable writers and speakers to make causal connections between old and new systems? What discursive features enable agents to destabilize and delegitimate existing social systems? What discursive features restructure and restabilize systems? As the title of this book suggests, I have framed these questions as problematics of the 'technology' of change by which I mean the techniques and functions (both material and discursive) that enable change to occur. This then is the story that the next chapters will examine and interrogate: how did OIT use the conditions of a failed technology implementation to change its function and role within Northeastern's social hierarchy? How did this group of people leverage a failed technology into a newly empowered social agency?

4 Discourse, stability and change

> The idea of a single break suddenly,
> at a given moment, dividing all discursive formations,
> interrupting them in a single moment and reconstituting them
> in accordance with the same rules – such an idea cannot be sustained
> (Michel Foucault 1972, p. 175)

Chapters 4 and 5 build on the theory and narrative of the previous chapters to present a close analysis of the texts used to construct social change during the WorkWare campaign. Chapter 4 initiates by reviewing Michel Foucault's ambition for a study of change that did not reinforce succession or chronology but attempted to situate change within discourse processes. Such a project may seem paradoxical but not when it is oriented to examine the conditions that enable change to happen. My own analysis begins with the study of clausal conditions and the interplay of dynamic and synoptic clauses as indicative structures of discursive stability during change. After considering clausal dynamics, the chapter examines methods for redefining subjects through copula verbs and subject complements (predicate nouns). The chapter concludes by assessing the function of determiners, given–new relations, and theme–rheme juxtapositions. Whereas the purpose of examining dynamic and synoptic clauses is to examine the conditions that enable (or are constituted by) change, the study of micro discursive features is designed to attempt to trace the activity of change at the grammatical level.

Stability, disruption, order: studying change through discourse

The problems of stability and disruption, order and variability, fixity and transgression permeate Foucault's (1972) *Archaeology of Knowledge*. Yet, for Foucault, these tensions are not so much problems as an entrance to characterizing change and transformation as a discursive process. These are moments, he wrote, 'in that white, paradoxically atemporal crack in which one sudden formulation replaces another' (Foucault 1972, p. 166). His point here is that the discourse formations that evoke change do not necessitate temporal succession, chronology or sequence. Instead, change occurs as discourses of discontinuity, atemporality and rupture (Foucault 1972, p. 166).[1]

For some, Foucault's project may seem paradoxical, or at least inconsistent. Especially when he claimed that his inquiry into change seeks 'general rules that are uniformly valid, in the same way, and at every point in time' (Foucault 1972, p. 166), yet he dismissed progressive and incremental sequences. He sought synchrony of discursive statements (Foucault 1972, p. 167), but was ambivalent about replication and the reproduction of the order and the activities of change. Yet, these claims may not be so paradoxical given that the subject (change) is always temporal and stochastic. The purpose here is not to outline a rational model for managing change or a framework to describe the reconstitution that occurs after change. Instead, the object of analysis is discourse. For example, locating identical discourses that reoccur at temporal intervals, but not as linked, progressive or incremental stages, suggests a different relationship between discourse and social transformation.[2] As David Macey wrote, 'Discourse is not to be interpreted in such a way as to reveal a history of the referent – of, that is, an object existing outside or prior to discourse' (Macey 1993, p. 202). Instead, discourse prefigures change. In other words, it establishes the conditions by which change may occur.

This emphasis on discourse signals an important break from studies of how to best deal with (or institute) the consequence of change. It is a different task to describe change itself, to articulate the brush strokes that create discontinuity. Concepts like managing change, instituting change and the various frameworks designed to map change are necessarily about resolving change: anticipating a consistent description of a process of change, prescribing how to effectively get workers/managers/clients to change, constructing a framework to describe how change takes place. Such projects rarely examine the conditions that enabled or created difference. As Foucault wrote:

> it is not enough simply to indicate changes, and to relate them immediately to the theological, aesthetic model of creation (with its transcendence, with all its originalities and inventions), or to the psychological model of the act of consciousness (with its previous obscurity, its anticipations, its favourable circumstances, its powers of restoration), or to the biological model of evolution.
>
> (Foucault 1972, p. 172)

In a twenty-first-century context, we can add to this list the structural analysis of organizations (in service of free market economics, maximizing shareholder value and industrial, supply chain efficiencies). There exists a justifiable pragmatism in such studies as they propose ways to understand, manage or predict a post-change resolution. But, in such cases, the event of change itself is taken for granted and rarely considered as an object or interest. As Foucault wrote, 'change' becomes 'both a general container for all events and the abstract principle of their succession' (Foucault 1972, p. 172).

As noted in Chapter 1, in addition to the more regularly accounted propositional content of texts, meaning is also created by textual (grammatical and organizational construction) and interpersonal (orientation and relationships of writers, readers, audiences) discursive functions. Even when propositional content may suggest a certain degree of stability, other features may betray the relative confidence a writer has in the claims being made or the level of fixity a proposition may have within an assumed context. As Halliday and Martin have argued, language is not passively reflecting some pre-existing structures; on the contrary, it is actively engaged in bringing such structures into being' (Halliday and Martin 1993, p. 8). Critical readers may anticipate or expect rhetorical manipulation within the propositional or interpersonal function of texts; yet few people will recognize the ways specific choices, conclusions and opportunities are prefigured within the textual function. As a dynamic situated in a moment between fixity and transgression, the moment of change must be realized and evoked within the textual function in addition to interpersonal and propositional functions.

Stability: clausal dynamics

In a very basic way, verbs and adverbs have been associated with dynamic meaning, in other words, meaning that is transitory, temporary or embedded within action or activity (Quirk *et al.* 1985, p. 75). Dynamic meaning has been contrasted with stative meaning which is associated with noun and adjective functions. Halliday and Martin have described these two textual forms of cohesion *dynamic* and *synoptic* (from stative) *orientations,* claiming that they can function as indicators of textual stability. Halliday and Martin wrote that looking at system from a synoptic perspective 'informs our conception of metastability' because 'it means taking synchronic snapshots of the meaning potential we construe as immanent' (Halliday and Martin 1993, p. 49). On the other hand, they argued that 'looking dynamically at system means looking at semiotic evolution, with evolution interpretable as expending meaning potential' (Halliday and Martin 1993, p. 49). Whereas synoptic systems demonstrate stability, without a dynamic perspective, 'it would be impossible for us to appreciate how a system can change' (Halliday and Martin 1993, p. 49).

Operationally (and analytically), the dynamic style is articulated by sentences with multiple clauses and embedded phrases. Vande Kopple (1998, 2002) has noted that Halliday's model works as a continuum of style with dynamic and synoptic articulating opposing poles. Vande Kopple used the continuum to document the interplay of dynamic and synoptic styles of articles written in the physics journal *Physical Review* from 1893 when the journal was founded to 1980. He argued that as the discipline matured, physics articles lost their dynamic orientation and became more closely associated with the synoptic style. Vande Kopple explained that the dynamic style is associated with 'spontaneous and unselfconscious speech' (Vande Kopple 2002, p. 192) and its 'chief characteristic of the dynamic style is grammatical or clausal intricacy' (p. 192). In other words, dynamic sentences include numerous clauses in both dependent and independent relationships with the principal subject-verb clause.

These multiple clauses are functional elements of sentences and in most cases are not obligatory. Specifically, Quirk *et al.* noted that adverbials are 'the most peripheral element' of a sentence (Quirk *et al.* 1985, p. 50), meaning that both formally and functionally, the adverbial can occur initially (before the subject), medially (in the middle between subject and verb) or finally (after the verb). A sentence (clause) may contain any number of adverbials or none at all. These characteristics position adverbial clauses as both formal and functional constituents of choice and not obligation. The same argument can be made for adverbial and prepositional phrases, which may add to the functional dexterity of a sentence (and obviously enhance interpersonal and propositional meaning) but are formally not obligatory elements of the sentence. As constituents that modify nouns, verbs and adjectives, prepositional phrases describe relationships between and among elements within a sentence or between sentences (Cook and Suter 1980, p. 201). Adverb phrases are noted by their mobility and optionality and their easy substitution by noun or prepositional phrase (p. 48). Quirk *et al.* used the example,

	just recently.
the weather has been very cold	*this month.*
	during the past week.

(Quirk *et al.* 1985, p. 48)

Yet, though all three phrases (noun, adverb, prepositional) are mobile, not all are replaceable. Quirk *et al.* demonstrated

This month has been very cold.
**Just recently* has been very cold.
**During the past week* has been very cold.

(Quirk *et al.* 1985, p. 48)

This simple substitution signals the obligatory nature of the noun constituent and the mobile, optional nature of the adverbial and prepositional phrase.

Multiple clauses, and to a lesser degree phrases, constitute the dynamic style and thereby suggest more unstable meanings because of their elective function. Stability here does not refer to truth-value or the accuracy of the specific claim being made. Instead, stability is a rough estimation of the confidence writers (or speakers) have in their ability to make a claim. Stability refers to the network environment into which a text is situated, the amount of lexical work a speaker or writer must accomplish to orient the audience to a claim, the amount of background information needed to support meaning, and other contextual issues that impede understanding. As claims require further and further modification (or explanation) they become more and more unstable. As such, the dynamic style is indicative of a changing, unstable network.

Typically, embedded clauses are the preferred linguistic device for this modification. Writers who have little confidence in their audience's level of comprehension will employ more clauses to properly elaborate and situate their claims, thereby creating a more dynamic style. Similarly, writers who anticipate significant resistance to a claim may choose to include clausal constructions to forestall opposing arguments. Embedded phrases enact a similar pattern within texts; however, since phrases enforce fewer demands on the sentence or its context their association with the dynamic style is not as overt as clausal constructions. At the same time, a text with multiple modifying phrases signals a less stable network given the writer's choice to include such modification.

Examining the clausal constructions of Northeastern Tech's Office of Information Technology's (OIT) emails throughout the WorkWare implementation demonstrates well how the operation of synoptic and dynamic styles constitutes and reflects a changing social network. As I will suggest, we can denote the way the social network (the conditions that enable change) is both reflected in and constituted by these clausal dynamics during and across the process of change. Figure 4.1 displays the sentence/clause ratio of each of the seven emails sent by OIT during the implementation campaign. For example, there are 28 sentences in email 1 and 68 clauses. This represents a 1:2.4 sentence clause ratio, which is higher than the ratio Vande Kopple cited for both early (1:1.82) and later (1:1.53) articles in his *Physical Review* data set. On average, the OIT emails are more heavily claused (1:2.14) than Vande Kopple's data, with a small decline in the number of clauses in emails 2, 3 and 7. Figure 4.1 does not include phrases.

Looking specifically at email 1, which initiated Northeastern Tech's WorkWare campaign, the first sentence begins with a subordinate clause (also called a 'reason' or 'reportage' clause [Quirk *et al.* 1985, p. 1105]), 'As President [...] mentioned in his correspondence to the campus on 24 July'. Here, the theme is announced in the left-most constituent of the sentence (President) and the conjunction 'as' combines both reason (because) and temporality (when) to state what is assumed to be a known fact. However, the clause is ambiguous as it creates a tension between power (*because* the president announced it) and time (*when* the president announced). The verity of the information in the main clause ([University] has entered) is dependent on the fidelity of the reportage clause. However, this ambiguity is not resolved. Instead, the email and (the implementation process itself) continues to leverage between the two interpretations of power and time.

Throughout email 1, when a claim addresses users' perceptions about the new software, reassures users, or provides reasons for using the new software, the relevant sentences contain multiple dependent clauses.

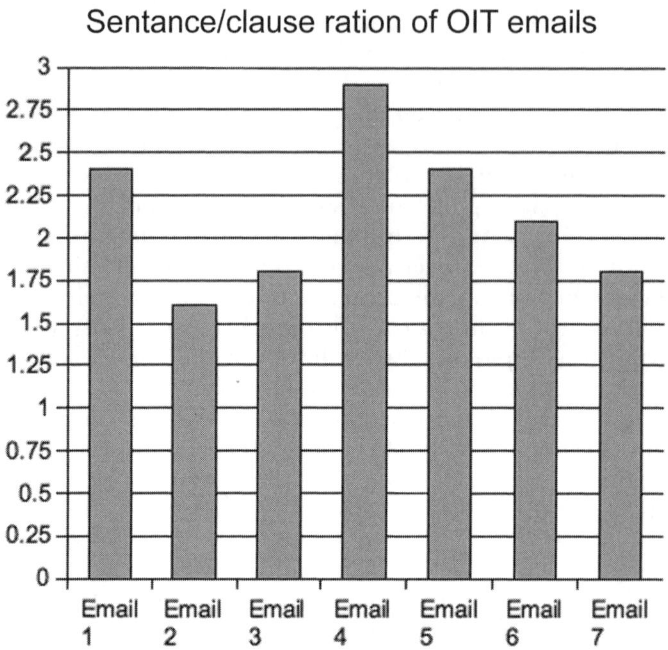

Figure 4.1 Sentence/clause ratio of OIT implementation emails

When a claim addresses facts such as the implementation time line or the specific features of the software, the sentence only contains one main clause (or the sentence is composed of two independent clauses joined by a conjunction). For example, the following sentence discusses user concerns and contains one main clause plus five dependent clauses:

> The purpose of this communication is to present | the key issues associated with the implementation | and ask you not to worry | if you don't understand | all the terms you will hear over the next few months. (Email 1)

Similarly, the next sentence asks users to endorse OIT's benevolent role in the implementation (one main plus five dependent clauses):

> If you remember only one thing from this letter | please let it be | that the Office of Information Technology is committed | to helping you use this tool | to assist in accomplishing your daily activities. (Email 1)

In contrast, the next example describes the features of the new software and is constructed by one clause (with an initial locative (adverbial) phrase and the main clause):

> In addition to traditional email and messaging, [WorkWare] | will enable integrated calendaring, document management and work flow, instant messaging and collaboration software, and an electronic workplace. (Email 1)

In the case of the more heavily claused sentences, key information is often located outside of the main clause. Email 1 employs 28 sentences and just over half (15) use complements. Cook and Suter identified two types of complements. They can operate as adjectives in a verb phrase that modify either the subject or the direct object of the sentence (predicate adjectives). Or, they can occur as noun phrases in a verb phrase that describe or redefine the subject or object (predicate nouns). These second redefining complements typically occur with the copula verb (is). In these cases, they are called 'subject complements' (Cook and Suter 1980, pp. 181, 183). Quirk *et al.* noted that in some cases, complements can be obligatory because they usefully complete the verb and supply vital meaning to the sentence (Quirk *et al.* 1985, p. 55):

> He stayed *immobile.*
> They kept him *at home.*
> He put down *the knife.*

In other cases, complements construct a more dynamic orientation because they are voluntary constituents. Complements are useful additions to the sentence as they can redefine the subject or complete the meaning established in the noun phrase. The following sentence from email 1 uses four clauses and two complements:

> We are *fully aware* | that many people have email addresses | that they do not want to change | and OIT is *more than willing* to accommodate this. (Email 1)

This example deals with users' perceptions about the new technology and potential concerns users may have about their new email accounts. Given the different rhetorical options available to the writer, the sentence remains dynamic. Yet, the inclusion of complements appears to begin to stabilize the context for OIT. For example, a more synoptic style could have produced:

> People do not wish to change their email addresses and OIT will accommodate this.

Or,

> OIT will not change your email address.

Whereas the above example deals with users' preferences (a highly dynamic context), this next example establishes the timeline for the WorkWare implementation. This sentence employs three independent clauses and three complements:

> This exciting project is *divided into two phases,* | phase I being *the WorkWare roll-out,* and | phase II being the *implementation of WorkWare tools.* (Email 1)

The independent clauses suggest that OIT is more confident in their time line, which at this point in the implementation they seemed to control. The complements neatly redefine the nouns: project —> divided; phase I ———> roll-out; phase II ———> tools.

In discourses of change, it is not surprising to find numerous complement constructions as these grammatical devices enable the writer to layer the text with correlating subjects and modifiers (e.g. we are aware; OIT is willing; project is divided). Of the 15 complements used in email 1, nine were used in multiple-claused (dynamic) sentences. Four were used in sentences with more than one independent clause joined by a conjunction. Only two were used in single-claused sentences.

Chapter 3 established Northeastern Tech's WorkWare implementation narrative. This was not a linear success story but involved considerable loss of trust and confidence as the software failed and constituent groups turned their backs on the implementation team and OIT. Recall that in response, the implementation team decided that campus users would not be required to adopt the WorkWare software. However, campus users were required to switch to a new hardware server which would route email to each individual client.

Using an approximate, qualitative narrative line, Figure 4.2 traces the level of contextual stability during the WorkWare implementation (the confidence writers had in the campaign) and correlates this against the clause/sentence ratio documents in Figure 4.1. In other words, the graph documents the grammatical use of clauses (dynamic style) in the emails and the progress and success of the implementation campaign itself – in effect correlating the discursive aspects of change with the material occurrences happening during the process.

As confidence in the implementation faded, the emails became more highly claused, even though the content of these messages did not overtly betray this situational instability. Quantitatively the sentence/clause ratio is expressed in Table 4.1.

Table 4.1 Sentence/clause ratio across implementation emails

Email 1	Email 2	Email 3	Email 4	Email 5	Email 6	Email 7
1:2.3	1:1.6	1:1.8	1:2.9	1:2.4	1:2.1	1:1.8

For example, email 2, written in August when Martin was still quite optimistic about the implementation, contains several prepositional phrases but few embedded clauses. Sentence 1, 'Northeastern is moving forward with the implementation of WorkWare' employs the copula (is) and two prepositional phrases, 'with the implementation' and 'of WorkWare'. Sentence 2 begins with a locative, 'over the next several weeks' indicating a degree of predictive uncertainty, but continues with the relatively stable construction, 'we will begin moving users'. The sentence

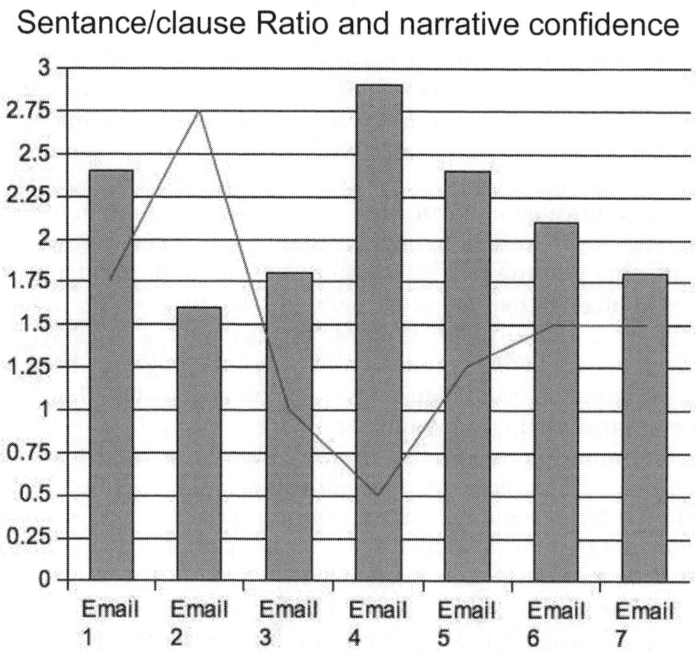

Figure 4.2 Sentence/clause ratio and narrative confidence

then ends appropriately with two prepositional phrases, 'from the current email and onto the WorkWare mail platform'.

These constructions can be contrasted with sentence 5 in email 3, written as Martin and his staff began to realize that WorkWare was not operational and that the campus was starting to raise concerns about the new product, the reliability of OIT's service, and the wisdom of this campaign. This sentence asks users who have not yet switched to Workware to contact OIT. The request was placed directly before the new narrative Martin designed which claimed that the university was testing WorkWare for its corporate partner.

> To help us in this effort, | please contact Kendra Smith <ksmith@university.edu>, | if you have not done so already, |to confirm that you wish to be migrated now. (Email 3)

Though rather short, the sentence begins with a nonfinite clause 'to help us in this effort' which precedes the main clause 'please contact Kendra Smith'. Two clauses and a phrase then follow: the clause 'if you have not done so already': the verb phrase 'to confirm' and the clause 'that you wish to be migrated now', which completes the sentence.

Email 4, sent in early November, emerged from the lowest point of the WorkWare campaign. It is the site of Martin's concession, 'If, however, you plan to remain with your current email client, we still need to move your mail service to the new servers WWS1 and WWS2, the servers used by WorkWare', which employs three clauses in addition to the main clause, 'we still need'. The email also enacts the third, final and minimal obligation, which is to simply move campus email to new servers. The instability demonstrated in email 4 is doubly layered as it not only emerges from the contextual instability of the failing implementation campaign, but it also emerges from the instability of the implementation team themselves changing their objectives and expectations in the middle of the project. Thus, the dynamic style evident in email 4 (1:2.9 sentence/clause ratio) is evidence of some instability in the progress of the WorkWare implementation campaign and it signals a possible change from within the context of the campaign itself.

Email 5 retains the relatively dynamic orientation (1:2.4 sentence/clause ratio) as the implementation campaign team struggled to regain their authority and credibility across campus. Recall that this email was written on the same day as email 4. The external context for the WorkWare implementation remains unstable. Here, the 'request' message, 'If you have not been migrated and would like to be, please call the helpdesk (x1234) or Kendra Smith (x4321), perhaps the most stable claim the team can make in this context, contains three clauses: 'If you have not been migrated'; '[if you] would like to be [migrated]' and 'please call x1234 or Kendra Smith'. The message to non-WorkWare users is also written in the dynamic style,

> The current mail server in use by [University] is over four years old | (that's 80 in computer years) | and is starting to show its age | – especially with the increase in email traffic over the last several years. (Email 5)

Despite the state of the implementation at this point, the WorkWare team did begin to assert some stability toward the conclusion of the email. The sentences following the example above suggest an increasing degree of stability as they attempted to convince hold-outs to switch to the new servers. The paragraph occurs at the end of the email and after making the case for the new servers (above) evokes a more stable orientation to explain how to shift to the new servers.

Recall, by this point, that switching to the new servers had become the principal objective of the WorkWare campaign. This switch, unlike the adoption of the WorkWare software, was necessary because the current, existing servers were going to be physically turned off. Though not the synoptic style, this technical certainty appears to help stabilize the next sentences,

> This process simply changes | the location of where your mail is picked up and delivered. Therefore, users who choose this option | will still be able | to use

their current client. This process is painless, | and in most cases, will be totally transparent. (Email 5)

The sentences above (in order) are comprised of two clauses: three clauses (viewing 'to use their current client[5] as a nonfinite clause with implied subject), and two clauses (viewing 'will be totally transparent' as a dependent clause with implied subject). The dynamic context is highlighted again in the email's final sentence, a request to users, 'If you have specific questions, | please call the helpdesk, | and we will have | a technician work with you | to explain the changes' (five clauses, with implied subject '*technician* explain the changes').

As the narrative in the previous chapter noted, no other emails were sent until 6 and 17 December. These final two emails used fewer clauses to re-establish the context of the implementation and provide an ultimatum to campus users. Email 6 establishes the context with a single-claused (multiple phrased) sentence, 'Over the past several months, the Office of Information Technology (OIT) has been migrating the email accounts of university email users over to the new WorkWare servers'. Sentences 2 and 3 are similarly synoptic, 'Approximately 70 users have not yet been migrated'. 'As of 18 December 2002, all campus mail will be redirected to the new servers'. This synoptic orientation emerges from a context in which OIT was secure in their own process and the inevitability of the switch to the new servers. Now that using WorkWare was no longer an issue, OIT's only objective was to install and operate the new servers. Yet, OIT makes one further request to the campus to cooperate with the switch to the new servers.

As the narrative noted, the intent of this email was to make sure no one unexpectedly lost their email. However, Martin and his team claimed that they did not know who remained unaware of the change and who remain belligerently opposed. The final invitation is thus appropriately dynamic, 'Regardless of the email you are using, if you have not made arrangements to have your mail service cut over to the new system prior to this date, you will not receive any new mail after 18 December 2002'.[3]

The final email (email 7) used the opposite construction to email 6. In email 7, the first sentence is appropriately dynamic, 'If you have not responded to the requests OIT has made over the past five months regarding the changing to the new WorkWare email servers, you need to call them by 4:30 p.m. today'.[4] Then, as the change came to a close and the situation stabilized itself (at least technologically, and for Martin in the completion of the implementation phase of the project), the last four sentences are single claused, 'The old email servers will be turned off at 6:00 a.m. tomorrow, Wednesday 18 December.' 'No new mail will be directed to these machines after this time.' 'Please contact the helpdesk x1234 with any questions.' 'Thank you for your cooperation during this transition.' The completion of the WorkWare implementation was thus signalled grammatically by synoptic construction at the conclusion of the final email. At this point the end of the transition was a certainty and the old servers would not function past the end of the graveyard shift. Compliance rather than change has taken over as the predominant social action of email 7. Although the resolution reached was far from the anticipated changes signalled in email 1, the situation stabilized, for now.

Re-defining the subject through copula verbs and subject complements

As the above discussion demonstrates, sentence–clause ratio can indicate levels of contextual (in)stability within discourse formations thereby helping us to understand the conditions that make change possible. This indication can point to dynamic orientations, which suggest changing social networks, or synoptic orientations, which suggest stabilized contexts. In some cases, synoptic orientations point to forced contexts or the use of power within a social context. What these orientations do not show, though, are actual cohesive links created throughout a process of change – how change can be solidified from within the instability of a dynamic context. Successful change must transition from dynamic (unstable) to synoptic (stable) contexts. Such a transition can be marked by several cohesive features, but most directly by the copula verb (is/are) accompanied by a subject complement (predicate noun). For example,

> One of the many *benefits* of this partnership is *the implementation of the WorkWare application.* (Email 1)
>
> What is *WorkWare?* It is *much more than just a new email system.* (Email 1)
>
> *Your comments and questions* are *important* to CP corporation. (Email 3)
>
> The *process* is *painless.* (Email 5)

In this construction, the subject complement (noun phrase following copula verb) redefines the initial subject providing a simple transition from one concept to the other and over the length of the email from early mention, to redefinition, to instantiation. The copula effectively reconstructs the local context enabling the writer to build a 'new' from an existing 'old'.

> The first phase of the WorkWare migration is approaching completion. (Email 3)
>
> Training (and or/retraining) is available... (Email 5)

The structure can also be used with future (will) and conditional (would) forms of the verb. However, these denote degrees of uncertainty since they imply prediction and desire.

> If you would like to be migrated to the new WorkWare system. (Email 4)
>
> As of 18 December 2002, all campus mail will be redirected to the new servers. (Email 6)
>
> The old servers will be turned off at 6 a.m. tomorrow. (Email 7)

The important feature here is that the change has been instantiated grammatically by assigning a new definition to the subject through the subject complement. Such complements redefine the subject in a way that does not engage the claim semantically or through argument. Instead, the audience is expected to accept the new alignment. As such, the argument functions as a discourse technology because the claim is constituted mechanically (grammatically) rather than semantically. Certainly, the subject complement must make sense semantically. But when multiple meanings (resolutions) are available, the copula + subject complement is an efficient way to manage audience interpretation and stabilize a changing social network.

Determiners, given–new and theme–rheme juxtapositions

Lassen (2004) has shown a second more subtle method of achieving discursive cohesion across changing social networks by examining given and new patterns embedded within theme and rheme constructions. 'Given' refers to information the speaker or writer presumes the audience already knows (is familiar) while 'new' refers to information that is assumed to be new information. 'Theme' refers to the initial reference of a clause (generally what the clause is about), typically the semantic content before the verb. 'Rheme' typically refers to everything after the verb. Fairclough characterized theme as 'the text producer's point of departure in a clause' and as such, theme usually corresponds to given information (Fairclough 1992, p. 183). Similarly, the rheme generally corresponds to new information. However, before proceeding too quickly with meaning structures, it is useful to briefly review the ways linguists have described the flow of meaning within a clause. At a basic level, the distinction between definite and indefinite articles (the/a) provides a starting point for understanding information familiarity and introduction.

Quirk *et al.* described the definite article (the) as 'referring to something which can be identified uniquely in the contextual or general knowledge shared by speaker and hearer' (Quirk *et al.* 1985, p. 265). Brown and Yule (1983) noted that this observation about the definite article in English dates as far back as two centuries. They provided the following citation from Harris (1751):

> 'Tis here we shall discover the use of the two Articles (A) and (THE). (A) respects our *primary* Perception, and denotes Individuals as *unknown*; (THE) respects our *secondary* Perception, and denotes Individuals as *known* (Harris, 1751, 215–16, cited by Brown and Yule 1983, p. 170.)

The refers to a referent with which the speaker/writer presumes the audience is already familiar. However, the assumption of familiarity does imply a certain amount of estimation and can be used to impose the fiction of existing information onto an unsuspecting audience. For example, Quirk *et al.* noted:

> In practice, since a speaker cannot always be sure of the hearer's state of knowledge, use of *the* involves a certain amount of guesswork. In fact, in some cases the assumption of shared knowledge is a palpable fiction. Notices such as *Mind the step* and *Beware of the dog*, for example, generally do not assume that the reader was previously aware of the hazards in question
>
> (Quirk *et al.* 1985, p. 266)

Similarly, statements like 'the memo was very clear about the policy' do little to educate or assist an unknowing audience but instead can use the convenience (fiction) of existing information to shift relations of accountability.

Unlike the definite article 'the', the indefinite 'a' refers to new information, information that is assumed to be first mention. Quirk *et al.* presented the indefinite article (a/an) as 'notationally the "unmarked" article in the sense that it is used (for singular or count nouns) where the conditions for the use of *the* do not obtain' (Quirk *et al.* 1985, p. 272). In situations where the speaker/writer is unsure about the audience's relation to or familiarity with existing knowledge, the indefinite article introduces the object as unknown. Specifically, '*a*' is used when 'reference of X is not uniquely identifiable in the shared knowledge of the speaker and hearer' (Quirk *et al.* 1985, p. 272).

For example, in email 1, we can see a transition from unknown information to instantiated and assumed information,

> ... [Northeastern Tech] has entered *a strategic partnership* with the CP Corporation. One of the many benefits of *this partnership* is the implementation of the WorkWare application.
>
> The purpose of this communication is to present the key issues associated with *the implementation* ... (Email 1)

Notice that WorkWare is introduced to the campus as a known entity, *the WorkWare application.*

In addition to determiner function, English grammar reports assumed information in two other ways which linguists call given/new relationships and theme/rheme construction. Brown and Yule (1983) noted that Halliday's (1967) initial categorization of given and new structures in English was based on spoken language which typically carried intonation patterns (Brown and Yule 1983, pp. 154–7). Halliday argued that in spoken English speakers typically begin with given (assumed) information and then provide new information. The emergence of new information is typically announced through a unique intonation which marks the information as new. Although such intonation patterns are not available in written English, Brown and Yule wrote that in most cases this structure of information sequencing is carried over from speech (Brown and Yule 1983, p. 177).

Prince (1981) elaborated on this structure by introducing a fuller taxonomy of information sequencing, including referents that are 'New' (brand new, unused), 'Inferable' and 'Evoked' (situational, textual)

(Brown and Yule 1983, pp. 182–3). Brand new information is information not known in any way to the audience. These references are typically introduced with the indefinite article (a/an). Unused information may be known to both writer and audience but may not be 'in play' at the time of composition or utterance. Brown and Yule used the example, 'I saw your father yesterday' as an example (Brown and Yule 1983, p. 182). Inferables are referents that the writer assumes the audience will pick up given the context of a preceding statement. For example, 'stairs' would be inferable in a discussion about climbing a building, given that most people recognize the existence of stairs in buildings: 'I climbed the Empire State Building last week.' 'The stairs were crowded with tourists.' Although 'stairs' occupies the given structure, it does not need to be introduced to be comprehensible because we can infer its referent from the preceding sentence. Finally, Prince's taxonomy described evoked referents. These are situationally evoked, which means they are available in the physical context, or textually evoked, which means they have already been mentioned in the text (Brown and Yule 1983, pp. 182–3).

Lassen (2004), examining discourse designed to promote genetically modified foods, identified structural patterns in which themes were built from information previously introduced as rheme. This finding follows an expected process in which information is initially introduced as rheme (and new) and subsequently moved to theme (given) status. However, Lassen was critical of this process when ideological information was moved to given status without appropriate documentation or proof. In other words, the only rationale for accepting the claim as fact became its location as 'given' information. Note that this function is similar to the function of resituating the subject with the subject complement following the copula.

The transformative patterns identified by Lassen are apparent throughout the WorkWare data set, but to a lesser degree. To an extent, the implementation campaign shows a consistent and expected pattern in given/new relationships across the implementation period. Information about WorkWare and the implementation is introduced in Email 1 as new and then placed into given structures in subsequent sentences and emails. I have *italicized* given information and **bolded** new information.

> What is **WorkWare**? *It* is much more than just a new email system.
>
> In addition to traditional email and messaging, *WorkWare* will **enable integrated calendaring, document management and work flow, instant messaging and collaboration software, and an electronic workspace**.
>
> Rest assured, as *these services* are introduced, plenty of hands-on training and user forums will be available. (Email 1)
>
> The first phase of *the WorkWare migration* is approaching completion. (Email 3)

Across these clauses, we can see the WorkWare concept was introduced as new information and then transitioned into given status. Its status is also taken for granted with the use of the pronoun 'it' which requires the audience to denaturalize the reference (WorkWare). Next, the sentences introduce WorkWare features as new information and then transition these to given status (and pronoun) as 'these services'. Finally, the final sentence positions 'the WorkWare migration' as given information. It has become something the writers assume the audience recognizes as used (already familiar) information.

Determiner function and the creation of social power

However, the given/new transformation is also mediated by the difficult contextual situation that accompanied the WorkWare implementation. This more contradictory context can be seen by mapping patterns of determiner (a/the) use to the theme/rheme structures. These relations provide added complexity within the given/new relationships demonstrating how many of each email's new statements include statements of presumed (or forced) knowledge.

The WorkWare campaign initiates in email 1 with the anticipated placement of the president of the university in given (thematic) position and the WorkWare system in the new (rheme) position. However, the email employs determiners (a/the) which construct levels of prior knowledge and understanding about the topic. As noted above, sentence 1 uses the new referent 'a' with 'strategic relationship', demonstrating that this relationship is new information to the campus. However, the sentence uses 'the' with 'the CP Corporation' which assumes that readers are familiar with this specific company. This initial presumption of familiarity may be relatively minor. However, other determiners presume familiarity with the implementation process and thereby assume a more familiar and stable context for the WorkWare implementation.

The first mention of WorkWare occurs in sentence 2: 'One of *the* many benefits of *this* partnership is **the** implementation of **the** WorkWare application'. Sentence 2 takes for granted the existence, familiarity, and knowledge of 'the many benefits' the partnership will provide the university. And, although 'implementation' and 'WorkWare' are placed in rheme positions, suggesting they are new information, a degree of familiarity is suggested by their introduction as 'the implementation of the WorkWare application'. Such placement, mediated by a familiar determiner (the) creates a degree of ambiguity in the sentence and the introduction of WorkWare. One interpretation would suggest that this demonstrates OIT's familiarity with the software. After all, they have been working with the system for several months now. Yet regardless of OIT's familiarity (or perhaps because of it) the claim grammatically structures the intended change by constituting WorkWare as familiar.

Notice the distinct difference between a less familiar and less confident introduction of the application if the introduction was composed this way:

> A benefit of this partnership is an implementation of a new software application called WorkWare.

As I noted above, the first mention of WorkWare assumes familiarity with the application and the plans to implement the software. This familiarity can be contrasted with the way OIT introduces their user-support Web pages (also in Email 1) which were designed to assist users during the implementation process:

> There is also **a** very informative Web page that explains **the** project in more detail. (Email 1)

Here, the website is introduced as brand new information in both indefinite determiner function and rheme position. Notice that the project is reintroduced with 'the' again, assuming familiarity of the implementation. In this case, 'the project' works as an inferable since readers can interpret the reference. Yet, throughout the implementation, new information is consistently introduced with definite articles. In some cases, this appears to be designed to bring concepts to immediate familiarity and acceptance. However, in other cases, positioning familiar issues in the rheme creates some ambiguity and tension between the authority for the implementation (theme/given) and the implementation itself (rheme/new but inferred).

> [Northeastern Tech] is moving forward **with the implementation of WorkWare**. (Email 2)

The new but inferred information could be placed in the thematic position as is accomplished later in Email 3:

> *The first phase of the WorkWare migration* is approaching completion. (Email 3)

By placing information about OIT as familiar and given, the clauses construct a new social position for Martin's office. Not only is the office positioned as familiar but crucially the office's authority and agency is also positioned as familiar and given.

Rheme–theme transitions across emails

Although I have presented examples of rheme-to-theme transition within individual emails, there are fewer examples of rhemes transforming into themes in subsequent messages. Email 2 reintroduces the (then) predominant purposes of the technical change as new (but familiar) information: 'the implementation of WorkWare' and 'begin moving users from the current email and on to the WorkWare mail platform'. Oddly, in the majority of subsequent references in other emails, these two topics do not attain 'given' status but are consistently introduced in rheme structures.

Email 3 begins, somewhat promisingly, by placing the implementation boldly in given thematic status:

> The first phase of *the WorkWare migration* is approaching completion. (Email 3)

However, this is followed by two sentences that place users' compliance with the initiative in new (rheme) status:

> All buildings have been visited but **not all requesting faculty and staff have been migrated**.
>
> The last week of September and the first week in October have been reserved **for the migration of everyone who remains**. (Email 3)

This lack of successful thematic instantiation is likely due to the way the implementation began to stall by email 4 as fewer campus users switched to WorkWare and those who had switched began to move back to older systems. For example, the interplay of theme and rheme from these first two sentences of email 4 highlight the dysfunctional context at this time of the implementation.

> *The Office of Information Technology* requests that you contact us **with your email requirements**.
>
> *If you would like to be migrated to the new WorkWare system, if you would like WorkWare training, or if you would like to discuss your problems or concerns with WorkWare,* please contact Kendra Smith x4321 or Ksmith@university.edu. (Email 4)

These themes present authority, uncertainty and OIT's agency rather than transitioned rhemes (new information) from previous emails about the implementation. Note that no rhemes from prior emails have transferred to thematic position in any clause in email 4. Instead, email 4 establishes a completely new set of rhemes. Later sentences attempt to restabilize the argument for switching to WorkWare, through direct rheme-to-theme transitions. These new rhemes introduce the radical new information that users do not need to use the WorkWare software.

> If, however, you plan to remain with your current e-mail client **we still need to move your mail service to the new servers WWS1 and WWS2, the servers used for WorkWare**.
>
> *This* is necessary because the current mail server in use by [Northeastern] is over four years old and is starting to show its age, especially with the increase in email traffic over the last several years.
>
> *The process* simply changes the location of where your mail is picked up and delivered.
>
> *This* will require a technican … (Email 4)

Email 4, which appeared at the most dysfunctional time during the implementation, demonstrates, through the given and new structures, the tensions at the moment in the implementation. In sentence 1, the Office of Information Technology is placed in given and thematic position, while user input (contact) is new information. Similarly, in each clause that makes up sentence 2, 'if' is the theme, while user needs are positioned as new information. Placing the user needs in the rheme suggests that OIT does not fully know why people are resisting WorkWare. This structure also suggests that OIT cannot assume or control the feedback they are receiving from the campus community.

The given/new dynamic is most functionally apparent in the placement of 'remain with your current email client' in the rheme (new) position in this sentence from email 4. This information subverts the entire WorkWare software implementation and changes the entire direction of the campaign. The sentence is composed of more than one clause which means that it has similar theme/rheme dynamics occurring across each individual clause.

> If, however, you plan **to remain with your current email client**
>
> *we* still need **to move your email to the new servers** WWS1 and WWS2, the servers used for WorkWare. (Email 4)

The concession, 'to remain with your current email client', is placed in new/rheme position of clause one (it is new information). OIT's agency is placed (asserted) in the given/thematic position of clause two (we). At this point in the campaign, as I will develop further in the next chapter, it is important to note how OIT's authority is becoming asserted (presented as given information). This is crucial for recognizing and understanding the constitution of the actual social change that is occurring during the WorkWare campaign.

Email 4 concludes by asserting OIT's commitment to the campus in thematic position, and again asking for user input in the rheme position:

> *In order to better serve the campus community, we* request **that you share the reasons you have for not moving to the new system**. (Email 4)

Email 5 continues this pattern in which the email asserts OIT's authority or status in the thematic position, which then relegates implementation information to the rheme. However, as the next example shows, an inferred rheme of email 3 has transitioned to the theme of email 5.

Email 3 (rheme):

> **not all requesting faculty and staff have been migrated**.

Email 5 (theme):

> All faculty and staff members who have expressed interest in moving to WorkWare have been migrated. (Email 5)

Yet, the themes of email 5 continue to highlight OIT's status and agency:

> *We* can also set up some of the preferences that you may have overlooked.
>
> *Our next step* will be to work with those who have not converted.
>
> *We* will contact you to discuss the concerns that you have.
>
> *Once we* have all these reasons documented, we will try to find solutions for each issue. (Email 5)

Other themes appear to have internalized campus problems and difficulties with WorkWare:

> *Training (and/or retraining)* is available on an individual basis or with small groups within departments.
>
> *Many of the difficulties users face* can be addressed during these sessions.
>
> *Other, more difficult problems* will be handled within a few days. (Email 5)

However, in discursive moves that reinforce that the technological change to WorkWare has yet to occur campus-wide, the email concludes with the (new) primary goals of the implementation still in rheme (new) position:

> *Our next step* is to work with **those who have not converted.**
>
> *We* will contact you to discuss **the concerns that you have**.
>
> *Additionally we* will be migrating **non-WorkWare users to the new server.**
>
> *We* need to move **all mail service to the new servers WWS1 and WWS2, the servers used for WorkWare**. (Email 5)

Although the implementation is situated in the rheme/new position, notice that OIT authority is placed firmly in given thematic position. The placement of the implementation as new information continues in email 6:

> Over the past several months, the Office of Information Technology (OIT) has been **migrating the email accounts of university email users to the new WorkWare servers.** (Email 6)
>
> Regardless of the email client you are using, if you have not made arrangement **to have your mail service cut over to the new system prior to this date, you will not receive any new mail after 18 December 2002**. (Email 6)
>
> We do not want **anyone to lose email service**. (Email 6)

Again, it is important to recognize that OIT's authority has migrated to the thematic position. Further, notice that in the second example (above) agency for not switching to the WorkWare system is assigned to the user in the theme position, 'Regardless of the email client you are using, if you have not made arrangement . . .'. I will take this up further and in more detail in the next chapter as we will examine the social change the technological implementation created at Northeastern Tech.

Finally, in email 7, the technical change is finally constituted by the given/thematic structure of the sentence:

> *The old servers* will be turned off at 6:00 a.m.
>
> *No new mail* will be directed to these machines after this time. (Email 7)

Discursive variables of the WorkWare technical change: implications and cautions

The purpose of this analysis, and its continuation in Chapter 5, is not to criticize OIT for the text of their implementation. Nor do I want to immediately suggest alternative ways for such processes. Instead, my purpose here has been to use these texts to describe and demonstrate the interconnectedness of grammatical and social structures in moments of change during the WorkWare campaign and to detail some of the micro-features of discourse technologies that emerged as prominent methods for instituting social change during this process.

As processes of change destabilize social structures, the way people construct, identify and refer to those structures in writing or speech must inevitably demonstrate aspects of dynamic and transient meaning. At the same time, as material and discursive forces restabilize contexts, realigning discordant narratives, texts will also coalesce into synoptic and familiar accounts dependent on the shared understanding of context and obligation. Because of this shared understanding, we readily identify such synoptic accounts as 'clear' and 'precise', by which we mean that the texts enact common groundings and understandings among readers and writers.

The obvious, unavoidable, yet unintended consequence of this study will be prescriptive directions to reduce the number of clauses in change messages, to avoid the dynamic style in total, and to replace 'a' with 'the' in successive change messages.[5] Such entrepreneurial activity has already nearly extinguished the utility of the passive voice, ensured that generations of young people will never begin a sentence with *and* or *because,* and curtailed the semicolon from productive use. A difficulty with descriptive studies is the ease with which they can be turned prescriptive because they do suggest a degree of prediction in their analysis. As an analytical device, sentence–clause ratio could be used to gauge the confidence writers may have in the context of their claims and knowing this beforehand may entice a writer to keep to a synoptic style.[5]

However, such attempts to short-circuit contextual referent may violate an audience's expectations for what linguists have called 'politeness' (Levinson 1983) or what Grice described as principles of cooperation in the use of language (Grice 1975, 1978). Chapter 6 revisits the concept of linguistic politeness to question whether technologically mediated contexts employ different assumptions about what constitutes polite and contextually appropriate discourse. But, before fully developing that discussion, it is important to review Grice's principles and detail the way a prescriptive grammar ('use the synoptic style') works against many expectations for what we have considered polite discourse. Levinson's summary of the principles of cooperation is a good place to initiate this explanation:

> *The cooperative principle*
> make your contribution such as is required, at the stage at which it occurs, by the accepted purpose or direction of the talk exchange in which you are engaged.

> *The maxim of Quality*
> try to make your contribution one that is true, specifically:
> (i) do not say what you believe to be false
> (ii) do not say that for which you lack adequate evidence
> *The maxim of Quantity*
> (i) make your contribution as informative as required for the current purposes of exchange
> (ii) do not make your contribution more informative than is required
> *The maxim of Relevance*
> make your contributions relevant
> *The maxim of Manner*
> be perspicuous, and specifically:
> (i) avoid obscurity
> (ii) avoid ambiguity
> (iii) be brief
> (iv) be orderly
>
> (Levinson 1983, p. 102)

The maxims may seem naive, if optimistic, for postmodern audiences inundated with advertising, political double speak, spin and the blog-sphere. Yet, despite their apparent dated idealism, the principles still help to interpret communicative intent (Levinson 1983, p. 131). Levinson wrote that 'for every kind of mutually assumed constraint on language usage, there will be a corresponding set of potential inferences that come about either from the speaker observing or flouting the constraint' (Levinson 1983, p. 132). In other words, when speakers choose to enforce synoptic interpretations of dysfunctional contextual arrangements, their violation of expected constraints (maxim of Quality) infers a deceptive communicative intent. Or, in the context of change, a sudden imposition of synoptic style forces closure and signals a desire to impose and enforce rather than change. Levinson set up such a situation in his discussion of Indirect Speech Acts (Levinson 1983, pp. 273–4). He asked why speakers often prefer a more contorted (and dynamic)

> I don't suppose that you would by any chance be able to lend me some cash, would you?

to the simpler and more straightforward (and synoptic)

> Please lend me some cash.

It is easy to see the first sentence as simply a wordy, self-deprecating distortion of the second. But, as Levinson noted, such an explanation does not explain the job that the dynamic style accomplishes or the preference speakers have for the dynamic in this specific example. Instead, Levinson argued that in this case, by deviating from the direct and concise, speakers 'communicate by conversational implicature' that considerations of politeness are taken into account in a difficult speech act (Levinson 1983, p. 274). We can add to Levinson's interpretation by noting that sentence 1 assumes and invokes a different contextual relationship between the speaker and audience than sentence 2. Whereas we may be able to invent situations where 'lend me some cash' may be more appropriate than 'I don't support that you would by any chance lend me some cash would you?', such situations pull the context between the speaker and audience into either a much more intimate scenario or they explode the context into an inappropriate, violent or criminal activity.

In a sort of round-about way, politeness indicators and our understanding of indirect speech acts limit the use of synoptic to stabilized contexts. Characterizing a changing context in synoptic style or using the synoptic style to impose on a dysfunctional context risks offending an audience and can be interpreted as an act of power. Thus, the synoptic proclamation in email 1 which introduces WorkWare is appropriately preceded by a more dynamic situating sentence:

> As President [...] mentioned in his correspondence to the campus on 24 July [Northeastern] has entered a strategic partnership with the CP Corporation. One of the many benefits of this partnership is the implementation of the WorkWare application. (Email 1)

Dynamic and synoptic orientations along with given and new structural relations signal the speaker/writer's understood relationship with the context in which an utterance/text occurs. It does not refer to perspectives of truth, intention or what the speaker/writer believes is true. It is an indication of what is understood to be possible to say at a specific moment and place. As such, it is a useful indicator of relative stability, instability, progress and resistance in a changing social network. And, in its violation, it can indicate activities of power, the repudiation of context and the imposition of force when and before change can take place. This discussion will be extended in the next chapter which considers the relationships and intersections between discourse and social networks during change. Chapter 6 will then further examine the function of politeness in change discourse and problematize the implications of this function within the context of discourse technologies.

5 Migrations and conversions

> The associated field is also made up of all the formulations to which the statement refers (implicitly or not), either by repeating them, modifying them, or adapting them, or by opposing them, or by commenting on them; there can be no statement that in one way or another does not reactualize others.
>
> (Michel Foucault 1972, p. 98)

> The work of contextualizing makes the connection between a context and a project completely unforeseeable ... The word 'conversion' has to be taken seriously, especially when it is uttered by an engineer who is also a theologian.
>
> (Bruno Latour 1996, p. 137)

> 'My name is Linus, and I am your God.'
>
> (Linus Torvalds, Linux Expo, Durham NC, 1998)

Chapter 5 continues the discourse-level analysis from the previous chapter. Whereas the previous chapter emphasized syntactic features, this chapter expands that analysis to include pragmatic features which influence social relations and ideology construction and promotion during the WorkWare change. The chapter begins by offering some additional information about the WorkWare team before moving to a discussion about modality, presupposition and the resolution of strategic ambiguity. This discussion describes how ideological beliefs can be discursively instantiated and disseminated throughout a social network. The chapter then explains how this process of modality and presupposition was implemented and became important for the WorkWare campaign. The chapter then switches the form of analysis to semantic features to examine the interpersonal forms of the implementation campaign and the various types and influences of intertextuality employed throughout the implementation texts. The chapter concludes by revisiting the principle of autonomous technology. By revisiting this concept here, I hope to demonstrate that the values and orientations that accompany technology are neither exclusive to nor necessitated by this specific medium. Instead, I argue that technology makes manifest those values, beliefs and orientations those enacting the technology have always already adopted.

Leadership discourse and WorkWare management

Perhaps it is disingenuous to reveal only now that Martin, Lyceum Corporation's designated officer to lead Northeastern's Office of Information Technology, is an enthusiastic student of organizational leadership. He initiated a weekly Bible study group within the Office of Information Technology (OIT) that emphasized Christian methods of leadership. He has regularly preached and conducted services at his church. And, for a while, he contemplated becoming a full-time pastor. Though he would not claim any correlation here, most of his senior leadership team share an evangelical orientation toward their personal and professional lives. Like the books aligned on his office desk, or the posters on the computer centre's walls, Martin's meetings easily and comfortably move through computing issues; biblical parables; and narratives and lessons about leadership, morality and conscience. Martin's faith is not imposing so much as it is present and available. He is as eager to talk Ernest Shackleton, Saint Paul's conversion on the road to Damascus, or customer service lessons from J. Willard Marriott, as he is policies for regulating file sharing in the dorms, computer security and virus protection, or parsing bandwidth. After several lengthy discussions with him I continue to wonder if he sees all of these issues as largely coterminous.[1]

Martin's discourse is deliberate and meaningful. As a leader, he has a keen focus on results. At times, his critics argue, at the expense of adequate consideration of how those results are achieved. Though, when tested, he turns unapologetically to the improved culture of OIT, his office's retention rate, staff satisfaction scores, successful reports from external consultants, and the various measuring devices he has used to assess his leadership. Martin is not George Babbitt (Lewis 1922). Although the stereotype is available, it is much too convenient and easy. He recognizes that, at times, he has not aptly negotiated the chasm between his third-party business unit and the larger academic community and culture within which his office is situated. Understanding the necessity of the degree in a higher educational context, several years ago he enrolled in a PhD programme with the intention to study organizational leadership. Predictably, he chose an online PhD programme with a for-profit corporate university. Some would see considerable irony in the way he promotes the convenience and efficiency of this medium to the president and provost of the quite traditional, residential institution that is his client and pays his bills. Yet, Martin is consistent in his values, motivations and actions. For him, distance education, online instruction and corporate models of profit-based service provision are necessary and inevitable steps for higher education.

Modality, presupposition and resolving strategic ambiguity

Though close in semantic meaning, 'necessary' and 'inevitable' are not the same thing. They evoke a gradient of meaning that can be dependent on degrees of time and personal or contextual need. Something may be inevitable without being necessary. Grammatically, parsing and conflating such distinctions is accomplished through a system of modality – the auxiliary verbs can, could, would, should, shall, may, might, will and must. Modality can be a key feature of destabilizing discourses (Faber 2003) and modals are unique from other verbs in that they express a degree of conditionality. For example, the phrase *Mary can swim across the river* expresses different meaning from *Mary swims across the river* or *Mary is swimming across the river.*

Functionally, modality can be characterized in two ways, intrinsic and extrinsic modality (Quirk *et al.* 1985, p. 219). Intrinsic modality (root modality) describes permission, obligation and volition. These concepts identify and complicate human control over specific events. Extrinsic (epistemic) modality refers to necessity, prediction or possibility. These concepts refer to events over which humans have no control. Thus, the modal 'can' in *Mary can swim across the river* could intrinsically refer to Mary's volition and willingness, her physical ability or the likelihood of her undertaking the swim. Extrinsically, 'can' could refer to the condition of the river, the current, depth or temperature issues, possibilities and forces of nature which Mary cannot control. In this way, modals are ambiguous because they fragment the rationale for activity by providing simultaneous, multiple explanations for activities. In some cases, intrinsic and extrinsic uses can be separated and in others modality can evoke both properties simultaneously (Papafragou 2000, p. 520).

Modality becomes an issue when writers confuse volition (human choice) with extrinsic (no choice) action. For example, the phrase '*With this new environment we will be moving toward a standard 1+7 naming convention so your user name will be consistent across the many systems we use at Northeastern Tech*' (Email 1) appears to evoke a necessity borne from technological change, obscuring the potential for volition and human choice. Yet, the requirement to use the naming convention was in actuality driven by human choice and motivation. At some point, someone decided to implement this specific aspect of the program. Yet this choice is obscured by the extrinsic interpretation of 'will' to mean necessity.[2]

We can also include the progressive aspect (is/are __ing) to the functional, strategic role played by modality (Faber 2003). Unlike traditional forms of modality, the progressive aspect focuses on time (they **are playing** hockey; she **is volunteering** for the hospital; corporate universities **are becoming** a competitor in the higher education marketplace). In these cases, the progressive aspect hedges the exact reference to when an activity started and ended. This makes it impossible to adequately evaluate the temporal duration of claims like 'corporate learning centres **are evolving** into universities', or 'education **is becoming a scarce commodity**'. We cannot tell when these actions initiated or how long they will endure.

Papafragou claimed that the ambiguity raised by modality is usually resolved during the process of oral comprehension (Papafragou 2000, p. 521). Yet, in written communication, this ambiguity cannot be immediately or readily resolved. Deliberately, modality can construct what could be termed 'strategic ambiguity', meaning an intentionally destabilizing claim that cannot be immediately resolved or solidified by available textual or contextual information. In addition, when accompanied by a presupposition, an assumed meaning embedded within an utterance, modality takes on a further strategic function (Faber 2003). Fairclough wrote about presuppositions in three ways, existential assumptions (assumptions about what exists), propositional assumptions (assumptions about what is, can be or will be) and value assumptions (what is good, desirable or bad) (Fairclough 2003, p. 55). In each case, a presupposition evokes a claim as shared understanding in a way that makes it difficult to challenge the verity of that claim. For example, the sentence below asserts the propositional fact of a global economy. The second overt claim is open to challenge but it is more difficult to challenge the presupposed claim (the existence of a global economy) because it requires an audience to break politeness conventions (interrupt the speaker, change topic).

> In a global economy, companies must focus on efficiency and profit maximization.

Strategic ambiguity is accomplished and resolved through discourse when speakers combine modality (including progressive aspect) and presupposition to assert an ideological claim that destabilizes a current context (Faber 2003). The model works because the modality creates a degree of ambiguity that the presupposition then interprets and resolves, creating an apparently rational stabilization. For example, take the following statement:

> As traditional universities become more and more detached from the real world, a corporate-based higher education programme will provide students with a competitive advantage.

The sentence employs the modal 'will' (will provide). The ambiguity here resides in extrinsic modality: 'will' can be interpreted as a prediction, a possibility or a necessity. The presupposition then provides an ideological key for resolving this strategic ambiguity. If the audience accepts the presupposition (that traditional universities have become detached), the modal becomes more ideologically invested and is interpreted as a necessity. In this way, the modal destabilizes the meaning context and the presupposition provides the ideological key (rationale) for restabilizing the context in a new way. Figure 5.1 outlines this pattern.

Modality	——>	Strategic ambiguity
Strategic ambiguity	——>	Presupposition
Presupposition	——>	Ideology
Ideology	——>	Textual stabilization

Figure 5.1 Modality and resolution of strategic ambiguity (Faber 2003, p. 406).

Modality and social change across the WorkWare implementation

Before fully examining the use of modality across the WorkWare campaign, it is important to note that modals are important to the campaign as strategic devices. They are not the predominant verb type used across the implementation documents. The discourse used throughout the implementation campaign primarily relies on the copula (is, are) with accompanying subject complements. Modal use is most frequent in emails 1, 2 and 5 with 'will' being the most widely used modal. Given that the email is an attempt to create a new technical and social system (a change discourse) we can predict a dynamic between the copula and the predictive modal. Cook and Suter wrote that 'will' is used for requests that do not require permission (Cook and Suter 1980, p. 92), for future states or actions, and for generalizations (Cook and Suter 1980, p. 95):

> Will you go to the store for me? (request)
> He will be taking the train tomorrow (future action)
> It will work on a Macintosh computer (generalization)

The modal 'will' expresses a degree of conditionality and uncertainty that must be resolved either materially, in the case of physical events, or discursively, in the case of ideological pronouncements/events. However, 'will' is ambiguous in most uses because we do not know which event is being enacted. As I will show below, the modality used across the WorkWare implementation campaign elicits future states and actions. These modals function initially to destabilize the existing technical and social context in which people on campus currently use their information technology. This destabilization occurs by blending intrinsic and extrinsic modality, suggesting that issues that are in people's control, issues of personal choice, are not controllable but actions of necessity and inevitability. Later uses of modality restabilize the social context by presuming technological inevitability and user compliance.

However, as I forecasted in Chapter 3, the WorkWare implementation narrative, in this new aggregation, the social change that emerges actually has little to do with WorkWare, with the choice to adopt software, or with complying with the new installation. Instead, the new aggregation inscribes OIT as a powerful social agent on the Northeastern campus. As the social context becomes more destabilized (as described in Chapter 4), Martin's discourse places responsibility for adopting the change onto campus users. However, the control over campus technology (regardless of which technology users choose) is retained by OIT. Once OIT owns all campus technology, users are free to adopt any technology they choose. Regardless of their choice, they are using OIT's technology and OIT emerges as a new and powerful agent on campus.

We can see this shift as part of the 'translation' or rearticulation that Latour cited as necessary for the successful technological project. Here we have a 'chain of translation' that propels OIT into a powerful campus agent from the ruins of the WorkWare implementation. As Latour wrote, translation requires a dynamic between a global and a local problem. At Northeastern we have several global problems: a fiscal necessity to find wealthy corporate sponsorship; a resurgent corporate interest in university education and its 'products' (research, students, facilities); an institutional predisposition toward technology and a desire to remain 'ahead of the technological curve'. These become translated into local solutions: a corporate–university partnership with CP Corporation; donated technology; technical solutions to administrative problems; WorkWare.

At the same time, Martin was aggregating his own global/local dialectic. He is an evangelical student of leadership at a prestigious technology management company (Lyceum) situated locally in a chaotic and discredited information technology context. Martin and his colleagues at Lyceum and the senior administrators at Northeastern and CP Corporation share a common interest in the global problems/opportunities for corporate capitalism on university campuses. Yet, the local, on the ground manifestation of these problems is quite different for Martin than for Northeastern. To continue to use Latour's logic, Northeastern's global/local dynamic is translated by an intermediary (Lyceum, an external, for-profit corporation, represented by Martin) who is motivated enough by the global problem to become interested in the local solution. Martin's chain of translation goes something like this: there is no solution to the organizational problems at Northeastern Tech without new technology; no new technology without corporate involvement; no corporate involvement without corporate ownership; no corporate ownership without Martin (Latour 1996, p. 33).

These chains of translation can be observed to be articulated through the sentence grammar of Martin's campus discourse. As the implementation progresses, Martin shifts his discourse from intrinsic (choice) to extrinsic (no choice) modality. Here, Martin asserts OIT's control of the campus IT infrastructure and he inscribes a new more powerful social role for OIT on campus as an entity that dictates what technologies people will use on campus and how people will use these technologies.

However, in those emails written as the WorkWare implementation began to falter, when it became apparent that WorkWare would not be fully adopted, Martin's discourse about WorkWare itself shifts to intrinsic (choice) modality. In this shift to intrinsic modality, Martin places responsibility for adopting WorkWare onto individual campus users, making the rejection of WorkWare an issue of user choice rather than an inability of his office to successfully implement the product. Finally, in the last two emails, which elide the WorkWare product altogether and simply dictate that campus users must direct their email software to the new servers, the modality is again extrinsic. This final shift to extrinsic modality again asserts that ultimately users do not have a choice but must use the new servers.

What I hope to point out in this discussion is not so much whether or not people could choose to adopt WorkWare, but the way modality enabled Martin to invoke both technical and social changes at two discrete levels. As described in Chapter 3, Martin was directed to implement the WorkWare software platform across the Northeastern campus. At the same time, as the new manager of a largely discredited and ineffective information technology office, he had hoped to use the WorkWare campaign to restore authority to his department. On one level, we can see a shift toward more conditional modality from emails 1 to 5 which documents a progressive failure to successfully implement the WorkWare platform. However, on another level, the same emails instantiate an extrinsic modality, a technical discourse that re-establishes the authority of Martin's department. These two discourses operated at the same time through the implementation campaign simultaneously failing and succeeding to achieve their strategic intent.

Table 5.1 summarizes the modals used in email 1 and their type (extrinsic/intrinsic). I have added to the table presuppositions asserted in the sentence which contains the modal and my own interpreted function of the modal.

The discourse of email 1 blends intrinsic and extrinsic modality, transforming actions of choice and volition into events of necessity and inevitability. There are few semantic arguments for adopting WorkWare presented in this email. Other than a technical description of the applications WorkWare offers, the content of this email offers no rationale for dispensing with current technologies in favour of WorkWare.

Instead, email 1 technologizes the argument, creating a discursive condition in which users have no choice but to use the new technology. The discursive technology is accomplished through the strategic ambiguity evoked in the tension between intrinsic and extrinsic modality. As noted above, the resolution of ambiguity is guided by presuppositions embedded within the sentence. For example, in sentence 3, if the presupposition is accepted (that users are worried about the new technology and don't fully understand the implementation), 'you will hear' evokes a lack of user control and ability to influence current implementation events contexts rather than a voluntary action to engage with the implementation process or be attuned to implementation news.

Martin's repetition and dependence on 'will' as a conduit from intrinsic to extrinsic modality is not a new or novel aspect of the discourse of technology. In a study of AT&T's 1993 advertising campaign based on the

Table 5.1 Modality in email 1

S#	*Use*	*Presupposition*	*Type*	*Function*
3	The purpose of this communication is to present the key issues associated with the implementation and to ask you not to worry if you don't understand all of the terms you **will** hear over the next few months	You worry about the implementation. You don't understand	Intrinsic/ extrinsic	Destabilize – removes agency/ control from user.
4	You **will** be receiving several email communications in the coming weeks that **will** give more definition to the new application.	The new software (WorkWare) is defined and well known	Intrinsic/ extrinsic	Destabilize – removes agency/ control from user.
7	In addition to traditional email and messaging WorkWare **will** enable integrated calendaring, document management and work flow, instant messaging and collaboration software, and an electronic workplace.	WorkWare enables email, messaging. People want these services (weak assumption)	Intrinsic/ extrinsic	Stabilizing through technological assertion: technology enables.
8	Rest assured, as these services are introduced, plenty of hands-on training and user forums **will** be available.	Services are inevitable. Users require training and assistance.	Extrinsic	Destabilize – users require training and assistance. Users have no choice about WorkWare.
13	As more features are added, WorkWare **will** be able to be used for document sharing; workflow management and establishing an electronic workplace that makes your documents available anywhere you have Internet access.	Added features, users want these features (weak)	Extrinsic	Destabilize – features are currently not available.

S#	*Use*	*Presupposition*	*Type*	*Function*
15	With WorkWare, whether you use Windows, Linux, Solaris, or Macintosh, **you'll** have access to your email and calendar, anywhere and anytime you have access to the Internet.	Users want access to these technologies	Extrinsic (but intrinsic implied)	Stabilizing – appearance of choice.
17	With this new environment we **will** be moving toward a standard 1+7 naming convention, so your user name **will** be consistent across the many systems we use at the University.	New environment	Extrinsic	Destabilize – user name no long in user's control.
18	In conjunction with the user name the new standard for email addresses **will** also be 1+7.	New standard for email addresses	Extrinsic	Destabilize – address out of individual control.
21	We **will** keep you informed as to when these changes **will** occur.	Changes occur	Extrinsic	Destabilize – out of individual control.
22	We realize that this announcement **will** generate many questions.	Users have questions	Extrinsic	Destabilize – out of user control.
23	To address these questions, and demonstrate the many features of this new software, we **will** conduct two open forums.	Address questions, demonstrate features, Software installed (weak)	Intrinsic (volition)	Destabilize – users have no authority.

similar slogan 'you will', Karla Kitalong argued that throughout the ad campaign the use of 'will' was used to situate and portray new communication technologies as devices for 'creating unprecedented, seemingly magical opportunities for consumers' (Kitalong 2000, p. 290). The television advertisements depicted small social events, such as a female executive working on a laptop while on vacation. In a voice over, the

AT&T advertisement asks viewers 'have you ever sent a fax from the beach?'. The answer to this and other, similar hypothetical situations, is the mantra 'you will'. Kitalong wrote that the phrase evokes a technological magic that, in the case of the fax-from-the-beach, enables users to, in her words, 'transcend space and time so as to simultaneously conduct business and interact with their families' (Kitalong 2000, p. 290).

Kitalong is correct that in the AT&T ads, and in OIT's implementation, modality enacts a transformation. The metaphorical catalyst for this transformation may simply be specific to one's disciplinary discourse or preferences for the technology presented. Humanists may prefer 'magic' while social scientists 'technology' to portray the transformation enacted by this discourse. In both cases, we see enacted Burke's *peithananke*, 'compulsion [enacted] under the guise of persuasion' (Burke 1950, p. 50). As such, these are arguments of power and strategy rather than deliberative processes of choice and volition. Simply because a technology makes something possible (sending a fax from the beach), the assumption becomes that using the technology is necessary and inevitable.

Negotiating technical necessity and personal choice

Subsequent WorkWare implementation emails continue to broker between technical necessity and personal choice. As the social and material conditions continue to support OIT, Martin is able to push the extrinsic interpretation of modality. However, as the context begins to destabilize, the extrinsic is replaced by modals of condition, choice and volition. Email 2, written while the campaign was still relatively confident, uses 'will' in two of its three sentences:

> 2. Over the next several weeks, we will begin moving users from the current email and onto the WorkWare mail platform.
>
> 3. We are following a schedule that will allow faculty and staff to be moved to the new system by 30 September 2002. (Email 2)

In Sentence 2, 'will' appears intrinsic as it identifies the technology office (OIT: 'we') as agents who are responsible for changing to the new platform. Sentence 3, however, shifts the actual authority for the implementation to an inanimate 'schedule' that technologizes when the actual change occurs. These modals allow OIT to demonstrate their control of the implementation process and at the same time evoke the necessity of moving to the new system.

Email 2 was written while the implementation was still relatively stable. Email 3, on the other hand, was written as the implementation began to fall apart. Understanding this material background, this email reads much like a final attempt to keep a system together. No other email throughout the campaign uses fewer modals (compared to other verbs). This email includes only one modal. However, its placement is key to understanding (and predicting) the social context that was about to derail the WorkWare implementation. Here, 'will' predicts a new rationalization and reformation of the campaign in one strategic utterance that acknowledges that WorkWare does not work, but unsuccessfully attempts to keep the implementation from fully unravelling:

> 5. The university will have a profound impact on the final WorkWare product. (Email 3)

This claim announced Martin's new implementation strategy and his attempt to convince an increasingly frustrated and sceptical campus to stick with the new software. Unlike the previous 'will' of Email 1, here users (the campus community) are brought along as co-agents to determine the exigency of the modal. The claim itself admits that WorkWare, in its current instantiation, is not 'final' but a work in progress. And, while Martin would like 'will' to indicate a necessity, most users probably read the note as a slight possibility.

Whereas 'will' predominates modal choice throughout the early implementation discourse, the more conditional and less assertive 'would' is used in email 4, as the implementation context further deteriorates. Here, the discourse turns to individual users' volition and places accountability for the successful implementation on both the users and OIT staff. Table 5.2 summarizes modal use in email 4 by sentence.

Although the email begins by shifting agency and accountability for adopting the software onto the individual user, the email concludes by again reasserting OIT's control of the process. Here, we see the divide initiated between implementing WorkWare and Martin's hope to reposition OIT's social presence on campus. Alternative wordings of email 4 might have kept the message directed toward users' preferences rather than OIT's needs. For example, substituting 'you may ask us to send a technician to assist you with this switch' for 'This will require a technician to visit your desktop' (sentence 6) would place agency for adopting the software and contacting a technician with the user. Yet, the re-occurrence of the extrinsic 'will' again removes such volition from the process. Email 4 may shift the obligation for choosing WorkWare onto individual campus users, but those users who choose WorkWare must also give up control over their desktop and their communications technology. As if to emphasize this separation between the user and the technology, OIT's message notes that the technician will not visit the user, but the user's computer ('to visit your desktop'). The user is again separated from the technology.

As noted above, here we see the first separation of social agendas in the implementation campaign. In response to the destabilizing WorkWare implementation, OIT appeared to shift accountability for adopting WorkWare onto the campus community. However, OIT understandably refuses to give up control over the information technology infrastructure – including the individual desktop computers people view(ed) as their own

Table 5.2 Modality in email 4

S#	*Use*	*Presupposition*	*Type*	*Function*
1	If you would like to be migrated to the new WorkWare system,	New Workware system	Intrinsic	Shift agency to users: volition/ obligation
1	If you would like WorkWare training	New WorkWare system; training	Intrinsic	Shift agency to users: volition/ obligation
1	Or if you would like to discuss your problems or concerns with WorkWare, please contact	WorkWare; people have problems, concerns with WorkWare	Intrinsic	Shift agency to users: volition/ obligation
6	This will require a technician to visit your desktop but it is [a] painless process ...	Implementation process (this) User assumes process is painful (weak)	Extrinsic	Sets conditions for participation
6	... and in most cases, will be totally transparent.	Implementation process	Extrinsic	Predictive: reasserts control
8	If you have specific questions pertaining to this switch, please call the helpdesk x1234 and we will have a technician work with you to explain the changes.	Concerns are technical (not policy or choice)	Extrinsic	Sets conditions for participation

personal machines. By email 4, users may choose not to use WorkWare. But, if they want to continue to use information technology resources on the university's campus, they must adhere to OIT's intended new social function on campus. As one social (and technical) change appears to fail, another much more powerful social change begins to materialize.

Email 5, released on the same day as email 4, echoes the shift in accountability iterated in the previous email and the 'move to power' also asserted there. The email begins with a more conciliatory tone using more intrinsic modals (can, may). This again provides users with the option to forgo the WorkWare software. However, like email 4, it concludes by re-asserting control through extrinsic (will) modality that substitutes the possibility of choice with technological determinism. Users must adopt the

Table 5.3 Modality in email 5

S#	*Use*	*Presupposition*	*Type*	*Function*
3	We will be happy to schedule a time that is most convenient for you.	N/A	Intrinsic	Volition: express willingness
6	It [training] can be very basic …		Intrinsic	Volition
6	… or it can be a time for you to discuss specific needs and problems concerning WorkWare.	You have needs and problems concerning WorkWare	Intrinsic	Volition
7	We can also set up some of the preferences you may have overlooked.	You overlooked preferences	Intrinsic	Volition
7	We can also set up some of the preferences you may have overlooked.	You overlooked preferences	Extrinsic	Possibility/ prediction
8	Many of the difficulties users face can be addressed during these sessions.	Users have faced difficulties	Intrinsic	Volition/ possibility
9	Other, more difficult problems will be handled within a few days	More difficult problems exist	Extrinsic	Asserting control and expertise
13	Our next step will be to work with those who have not converted.	Others have not switched to WorkWare	Extrinsic	Asserting control over those who have not switched
14	We will contact you to discuss the concerns that you have.	You have concerns	Extrinsic	Control
15	Once we have these reasons documented, we will try to find solutions for each issue.	Reasons for not switching exist	Extrinsic	Control
16	Additionally we will be migrating non-WorkWare users to the new server.	Non-WorkWare users; new server	Extrinsic	Control

S#	*Use*	*Presupposition*	*Type*	*Function*
20	Therefore, users who choose this option will still be able to use their current client	Users may choose not to use WorkWare	Extrinsic	Control
22	If you have specific questions, please call the helpdesk and we will have a technician work with you to explain the changes.	Changes are technical (not choice or policy)	Extrinsic	Control

new technological infrastructure OIT has installed. Table 5.3 summarizes modality in email 5.

The destabilized context – software that doesn't work, users switching back to previous platforms, university administrators reneging their support for WorkWare and, ultimately, Martin's decision to no longer require users to adopt WorkWare – becomes evident in the use of more conditional modality in email 5. The material context realistically no longer allows Martin to insist on full adoption. At the same time, given the problematic and contested history of campus information systems at the university, he must restore his office's credibility and its capability to operate information systems on campus. Although the use of WorkWare has now become an issue of volition (rather than necessity or obligation), the server substitution is Martin's final and only action remaining. He has little choice now but to insist that users adhere to his authority.

Since the new servers have already been acquired and Martin has the authority to install and use them for campus email, the argument to switch to the new services is entirely and solely material. On this issue, Martin is stable and confident. We can see this in the resumption of the extrinsic modal 'will' when Martin refers to people who are not using WorkWare (email 5, sentences 16–22). The extrinsic modal places control over the technological network back in Martin's hands, moving issues that have become volition (switching to WorkWare) to necessity (migrating accounts to the new servers).

The extrinsic necessity initiated in email 5 is resumed in emails 6 and 7. Email 6 begins in the past tense, establishing a common history and account of the technological and social change. (Note: the common history mentions only the servers not the software.) However, the email adopts the past perfect progressive, indicating an action that was started in the past and has continued to the present. While this may have been an odd strategic choice because it reveals that the implementation is not complete (that some users have still not complied), it does reflect the accuracy of the yet-to-be-stabilized social context.

> Over the past several months, the Office of Information Technology (OIT) **has been migrating** the email accounts of university email users over to the new WorkWare email servers. Approximately 70 users have not yet been migrated. (Email 6)

Using the past tense and writing 'the Office of Information Technology **has migrated** the email accounts of university email users to the new WorkWare email servers' would have constituted a more stable and complete context in the final stages of the implementation. But, the past tense (has migrated) would have also problematized the agency of those who had not migrated. It would appear that people had the power not to switch to WorkWare. The problem would have been even more evident if Martin had followed with, 'Approximately 70 users have not migrated'.

In its existing form, power remains unstable and negotiated, even here at the end of the implementation campaign. The verb structure 'has been migrating' makes it possible to claim that it is only a matter of time until the hold-outs start using WorkWare (Approximately 70 users have not yet migrated). To give users the accountability for not migrating would imply that this is something users could have done on their own – weakening OIT's claim to power. By retaining the accountability for the migration, OIT admits a degree of failure; they have not migrated these outstanding users. But, the present perfect progressive (have not yet been migrated) achieves Martin's social change in the relationship between the campus community and his office. It claims that these hold-outs will indeed migrate.

Martin's social change is solidified with the two extrinsic interpretations of the modal 'will' in email 6:

> As of 18 December 2002, all campus mail **will** be redirected to the new servers.

> Regardless of the email client you are using, if you have not made arrangements to have your mail service cut over to the new system prior to this date, you **will** not receive any new mail after 18 December 2002. (Email 6)

While choice (obligation) appears to be switched onto the user, control to affect the technology (and the user's email) again remains with OIT. Users are now responsible to make arrangements to maintain their email as OIT fully re-establishes their power to influence the technology. Despite that the technological changes directly originally associated with WorkWare did not occur, Martin was able to use the implementation to successfully claim authority and integrity for his new office and its role on campus. Martin was able to use the technological change to assert his own social change – a new campus information department that wielded considerable power across campus.

Thus, although the WorkWare software implementation campaign was not fully successful, Martin's own social/organizational change seems more stabilized and effective. Email 7 confirms this new social position as the email begins by situating accountability for the change fully on users.

Extrinsic modality then makes the ensuing technological change inevitable.

> If you have not responded to any of the requests OIT has made over the past five months regarding changing to the new WorkWare email servers, you need to call them by 4:30 p.m. today. The old email servers will be turned off at 6:00 am tomorrow, Wednesday 18 December. No new mail will be directed to these machines after this time. Please contact the helpdesk x1234 with any questions. Thank you for your cooperation during this transition. (Email 7)

'This transition' can be seen to be ambiguous as it may refer to the transition to the new servers, the problematic transition to WorkWare, or the transition that repositions OIT as a powerful agent on Northeastern Tech's campus.

A newly empowered OIT: interpersonal relationships

The strength of OIT's new position on campus is apparent in Martin's willingness and ability to technologize complicity in emails 6 and 7. Six months earlier, his social and material currency did not allow such a forceful discourse. Recall that the campaign itself was strategically announced as a presidential mandate. This cultural shift in power between Martin's office and the Northeastern Tech campus is also apparent in the interpersonal relationships enacted during the campaign. In email 1, Martin positions OIT as a service organization providing expertise for the user community. Email 1 states:

> [y]ou will be receiving several email communications in the coming weeks that will give more definition to the new application. If you remember only one thing from this letter please let it be that the Office of Information Technology is committed to helping you use this tool to assist in accomplishing your daily activities. (Email 1)

OIT asserts control of the implementation, yet there is a certain amount of benevolence inscribed here as Martin attempts to connect WorkWare to users' specific work needs. This benevolence seems to dissipate as the implementation context becomes more complicated. In addition, whereas emails 1 and 2 situated OIT as agents for the campus, email 3 introduces a slight differentiation in this relationship as the campus is positioned to be serving (or assisting) OIT. Note the differences in relationship embedded in these statements:

> The end result of this endeavour **is to provide you** with the latest technology to help you work more efficiently, and ultimately, provide our students with the best educational and research opportunities. (Email 1)

> Over the next few weeks, **we will begin moving** users from the current email and onto the WorkWare mail platform. (Email 2)

> The last week of September and the first week in October have been reserved for the migration of everyone who remains. T**o help us in this effort,** please contact Kendra Smith <ksmith@university.edu> if you have not done so already, to confirm that you wish to be migrated now. (Email 3)

Email 3 situates a new relationship between OIT and the campus. The email differentiates between campus users who 'have been migrated' and those who have not. Those who have not migrated are placed in the competing subject positions of either assisting or resisting OIT. This strategic move articulates the implementation in more personal terms than previous emails. In addition, the discourse addressed to those who have not yet 'migrated' offers fewer choices and options. Two weeks are set aside 'for the migration of everyone who remains'. Further, as email 3 introduces the campaign to assist CP Corporation with the ongoing development of WorkWare, OIT situates those who are using WorkWare as partners in an ongoing project. Those who are not using WorkWare are excluded from this partnership.

As the campaign intensifies its focus on those who are not using WorkWare, non-users are increasingly situated as marginal to the campus community. These are people who require special assistance, have special problems, or are difficult to work with. Email 4, which is specifically addressed to non-users, initiates and concludes with a direct challenge to these outsiders:

> The Office of Information Technology requests that you contact us with your email requirements. (Email 4)

> In order to better serve the campus community we request that you share the reasons you have for not moving to the new system. (Email 4)

Email 5 attempts to recreate a relationship with non-users (and to repair the relationship with those who adopted the program but are moving back to prior software). Here, again training is offered and OIT acknowledges that people have had problems with the WorkWare product. This good will is exhausted by emails 6 and 7 which treat non-users as antagonistic to OIT's agenda.

> ... if you have not made arrangements to have your email service cut over to the new system prior to this date, you will not receive any new mail after 18 December 2002. (Email 6)

> If you have not responded to any of the requests OIT has made over the past five months regarding changing to the new WorkWare servers, you need to call them by 4:30 p.m. today. (Email 7)

This antagonism is not difficult to understand, given the difficulties OIT endured throughout the implementation. Reflecting on this time of the implementation, OIT employees noted in interviews that they had little patience for people who refused to comply with their requests. As Abigail, an author of the early emails, noted to me, there was 'no empathy with the user resistance'. She and her colleagues 'couldn't really understand why people didn't want to switch' (interview notes).

Intertextuality and discursive motivations

The discourse of insiders and outsiders is embedded within a larger intertextual discourse Martin employed to characterize the implementation. Kendra claimed that the term 'migration' was used to characterize the implementation because it was 'a more technical term' to explain the switch. Though, in her mind, 'move' and 'migrate' were interchangeable, and she became tired of using the same word 'move' so she switched to 'migrate'. She thought that 'migrate' was not a word the general public often uses so it provided a 'more technical flair' to the campaign and 'downplayed the transition' (interview notes).

Kendra's account has some similarity with Martin's recollection of how the terms were chosen. However, it should be noted that the term 'moved' is not used in email 1. The email proposes a 'smooth migration of current email' (sentence 16). 'Migration' is preceded by 'implementation' (sentence 3), 'roll-out' (sentence 10), and 'introduction' (sentence 11). Email 2 uses 'implementation' and 'move'. Migration is stabilized as a concept in emails 3, 4, 5 and 6. In nine of ten uses of the term 'migrate' over these emails, OIT retains agency over the term:

> All faculty and staff members who have expressed interest in moving to WorkWare have been migrated. (Email 5)

In one use, users are credited for having migrated:

> Currently, over 400 faculty and staff members have migrated. (Email 5)

Emails 6 and 7 shift away from this vocabulary, using the terms 'redirected', 'cut over', 'changing', and 'transition'.

Those who do not adopt the new system 'remain' and do not 'move'. Perhaps most vividly, email 5 describes those who have not switched to WorkWare as those who have not 'converted' (sentence 13). To some degree, Martin's investment in the project and his frustration with non-users is apparent in this choice of vocabulary. In my interviews with him, Martin claimed that for him 'migration' refers to a process of switching from one platform to another or from one machine to another; 'implementation' means that nothing like this has been used before; 'conversion' refers to updating a similar application, from version 1 to version 2 for example. He argued that 'conversion' carries with it a 'backend' change that is not necessarily noticeable to users. However, given his religious background and his immersion in evangelical management literature, perhaps we need to pay more attention to Martin's use of the term 'convert'. 'The word "conversion" has to be taken seriously,' wrote Latour when his own research subject used the word, 'especially when it is uttered by an engineer who is also a theologian' (Latour 1996, p. 137).

Intertextuality, to cite Fairclough, is 'the property texts have of being full of snatches of other texts, which may be explicitly demarcated or merged in, and which the text may assimilate, contradict, ironically echo, and so forth' (Fairclough 1992, p. 84). Originally developed by Kristeva and inspired by Bakhtin's approach to textual study (Fairclough 1992, p. 101), the concept of intertextuality refers to the multiple instantiations of texts within other utterances. Fairclough wrote, 'all utterances are populated, and indeed constituted, by snatches of others' utterances, more or less explicit or complete' (Fairclough 1992, p. 102).

Thus, common American speech contains numerous intertextual references to sport (strike out, score, slam dunk), business speak borrows heavily from the military (plan of attack, barriers, mission critical) and other cultural phenomena (ball park, fast track, bread and butter). While these are obvious references, the point of intertextuality is that language necessarily carries other meanings, references and influences. Fairclough argued that intertextuality can be an important feature of social texts and texts of change because through these relationships authors and speakers can channel other discursive formations and align their text with the ideologies and power relations of another group.

To an extent, Lakoff and Johnson (1980) noticed this in their work on metaphor, citing examples of specific metaphors that alter interpretations of events (metaphors of war used to describe interpersonal disagreement). Intertextuality expands this awareness to larger discourses that are not necessarily metaphorical. In addition, certain forms of structural intertextuality can point to structural constraints on a discourse, generic characteristics or even modes of resistance and opposition. Fairclough (1992) characterized 'explicit' and 'implicit' intertextual relations (those overtly noted and other, more subtle forms of reference). He also referred to 'manifest intertextuality' as a way of marking texts that are explicitly made present in a discourse. Fairclough also coined the term 'interdiscursivity' to refer to discourse conventions, by which he meant the constitutive relations that go into textual production (Fairclough 1992, p. 104). For example, Fairclough referred to turn-taking patterns, politeness conventions and genre scripts as examples of interdiscursivity since these discourse patterns construct acceptable styles and requirements for discursive activity (Fairclough 1992, p. 124). A direct reference to an athletic event in the context of a business meeting, 'that proposal was not even in the ball park', serves as manifest intertextuality. Integrating the genre of a newspaper article into the design of a printed advertisement would serve as an example of interdiscursivity.

As noted above, intertextuality can be an important technology for social change. Intertextuality makes present an absent alternative narrative through semantic reference, symbolic use or even structural imitation. These references do not need to be intentional or even conscious. In fact, change often can be sparked by an unintentional or even unnoticed intertextual connection that destabilizes or presents another option within an existing framework. Intertextuality is also difficult to resist since the practice calls into being cultural references and interpretations that are difficult to resist or challenge because they occur at interpretive levels rather than overt semantic (content) levels. For example, using intertextual references to new capitalism to refer to education, 'student consumers demand greater access to their faculty', materializes interpersonal assumptions about students, faculty and the ideology of education. Like presuppositions, these assumptions cannot be easily resisted because they frame the conditions of discussion, establishing the legitimate vocabulary that can be used in the discussion.

Thus, once a specific discourse begins to take hold within a social context, it must be resisted at a semantic and symbolic level rather than at the level of claims and argument. Yet, to what extent do Martin's intertextual references to migration, moving, to people who 'remain' and to 'conversion' make present an absent narrative? To what extent do his staff and his leadership team recognize and identify with this absent-but-present discourse?

Angels and demons in the WorkWare implementation[3]

Twelve years out from Umberto Eco's perhaps tongue-in-cheek assessment that the 'Macintosh is Catholic and DOS is Protestant' (Eco 1994), religious discourse remains prominent in the aggregation that constitutes computer technology and technology management. BBC correspondent Stephen Evans provided another example in his review of the Macintosh iPod (21 April 2005). Evans wrote that entering the Apple store in Manhattan 'is to enter a temple'. He described the 'cult' of the 'Mac', suggests that PC users are being 'converted ... to the higher values of the Mac', and depicted 'Mac' users as 'the faithful'.

The aggregation of religious discourse with the discourse of technology achieves many common purposes. These are both discourses that establish insiders and outsiders based on ideological belief and what is largely faith-based practice. People must deliberately choose to belong in such communities; yet, those who do not belong are situated as having deliberately and overtly rejected the community. Similarly, for Martin, those people who do not migrate are excluded from the community and are discursively 'left behind'.

Kenneth Burke, writing about what he called the 'Rhetoric of religion', claimed that 'technologism is a religion' (Burke 1950, p. 170). He specifically cautioned:

> Insofar as 'technologism' is a 'religion' (and it is a 'religion' to the extent that technology is viewed as an intrinsic good, so that its underlying, unspoken assumption is: 'The more technology, the higher the culture'), we had better favor a calculus that keeps us always aware of technology's possible relation to theology's vast motivational Cathedrals.
>
> (Burke 1950, pp. 170–1)

Working from Burke's caution, Peter Case directly compared the discourse of modern management technology, specifically the motivational discourse of business process re-engineering, to overt religious discourse (Case 1999). Emerging from the field of information technology and information management, business process re-engineering is a distinctly technical (computer-based) intervention into management theory and practice. Grounding his study in rhetorical criticism, Case summarized, 'the motivational patterns that one finds in sacred discourse thus mirror the social motives of everyday organizational life' (Case 1999, p. 422). In his analysis, he showed that the re-engineering literature relied heavily on religious discourse not only for metaphor but also for essential concepts that motivate, structure and frame social and organizational relations within an organization.

Thus, business process re-engineering vows an edict of simplicity, 'to do more with less' (Case 1999, p. 425) and evokes a sense of guilt for the sins of prior management practices – especially for past 'paradigms' which have only led to cost over-runs and organization wide inefficiencies (Case 1999, p. 431). Business process re-engineering's proponents chastise specific idols and practices of past management theories. Case wrote, 'past mistakes and misfortunes are attributed to agents (Adam Smith) and conditions (function division of labour etc.) which business process re-engineering promised to purge' (Case 1999, p. 432). As a saviour, business process re-engineering offers redemption from management's past sins through purging bad habits and sinful excesses 'and through such purgation, salvation for the managerial masses becomes a seeming possibility' (Case 1999, p. 432). Finally, Case wrote, 'unbelievers must be "born again," however, in order to partake of the sacrament' (Case 1999, p. 432). And, of course, those who are born again, who confess 'to the historical failure of IT solutions' (Case 1999, p. 433) will 'convert' (Case 1999, p. 436) to the new management doctrine. If, as Eco postulated above, the discourse of Macintosh is Catholic and the discourse of the PC Protestant, it seems that the discourse of technology management is equivalent to another American invention, Evangelicalism.

In its more extreme forms, this combination of management, technology and religious discourse shares intertextual features with another evangelical discourse promoted by the Tim LaHaye and Jerry Jenkins apocalyptic *Left Behind* novel and film series. Based on a literal interpretation of the biblical book of Revelation, the series includes some 15 fictional novels, ten graphic novels, three military-based novels, 40 children's books, a feature-length film, plus boxed sets, calendars, greeting cards, audio books, video, films, music and a role-playing computer game (www.leftbehind.com/). The narratives in this corner of the evangelical world replay again and again events predicted in the biblical book Revelation, paying specific attention and detail to the 'rapture', an event when true believers are taken (migrated) to heaven, literally disappearing, leaving the rest of humanity behind to endure a wretched life on earth.

It is easy for such discursive speculation, admittedly, to get out of hand. Substitute 'technology' or 'management' for 'science', and Burke put this temptation well, writing, 'the resources of dialectics being what they are, one can make out a good case either for the thesis that modern science grew up outside the rationalizations of theology or for the thesis that modern science is but the translating of theology into a secular counterpart' (Burke 1950, p. 170). At the same time much of Martin's rhetorical project at Northeastern can be better understood when articulated through this evangelical framework. Martin's leadership is based on a firm faith in his personal characteristics as a leader and the intentional 'values-based' community he has built in his office. Martin anticipates acceptance because it is the right thing to do and he has a difficult time understanding why other people may not implicitly trust him. Yet, his communal, faith-based rhetoric runs contrary to the highly secular and individualist culture of the university where he works but is not employed. In this sense, Martin is situated in the secular university community but he is not 'of' that community.[4] Few professors and staff would have even recognized any intertextuality within his discourse. And those who might would find it at best confusing and at worst offensive.

Thus, although we can see in Martin's rhetoric intertextual links to what for him and his staff are known and valued discourses, there was little currency in this discourse for other users on campus. This could help to explain why this discourse was dropped at the end of the campaign in favour of the more pragmatic terms 'redirected', cut over', 'changing' and 'transition'. With or without its evangelical references, 'migrate' assumed a communal perspective and sense of agency that did not exist on the campus. It assumed a group mentality that is not shared by many academics and it presupposed a symbolic currency that was lost on most people.

Autonomous technology revisited: individual values made manifest through technology

The purpose of this chapter has been to describe and characterize the discursive formation of interpersonal relations during a process of change. Initially, the chapter examined the use of modality, showing the way the implementation documents operated between intrinsic and extrinsic modality, modality that offered users choices and responsibility and modality that necessitated complicity and compliance. OIT's implementation campaign often interpreted choices of volition as demands of necessity asserting that users were required to use WorkWare. This extrinsic use complemented the technocratic move Martin made in the early emails. This strategy combined the use of the WorkWare servers with the implementation of the WorkWare software – essentially making both projects one single implementation. However, as the material context began to destabilize, we saw greater use of implicit modality as Martin was less able to technologize the implementation.

Martin emerged from the destablized implementation context with two separate agendas, the continued, but delegitimized WorkWare implementation, and the larger project to gain credibility and authority to his office, OIT. The implementation offered Martin the opportunity to gain a more authoritative and empowered social position on campus.

To achieve this position, Martin did not need WorkWare to succeed so much as he needed to convince users that they were dependent on OIT for their information technology needs. This overt claim was announced early in the implementation (email 1) and it was supported discursively through to the end of the final email (email 7). Martin created this dependency materially, but also discursively, by insisting that users required OIT to enable their computer technology. The emails never gave users the agency to switch their own email service to the new servers. Instead, they required that users depend on OIT for this relatively small fix. Martin used implicit modality to give users the choice to use the WorkWare software, but he used extrinsic modality to constitute and assert his office's more powerful campus authority. In this dynamic between volition and necessity, choice and obligation, what became at stake was no longer the implementation of a software program but social and material position and legitimacy on the university campus. Here we can see actions that radically influenced this social culture but were never overtly stated or semantically communicated. Instead, through specific discursive activity, supported by material action, Martin was able to reconfigure his position on campus and create a network of power for his office.

Recall from the previous chapter that the themes from OIT's communication switched from WorkWare-specific themes to themes that constituted and asserted OIT's authority, legitimacy and campus power. Note these themes from email 5 (as examined in Chapter 4):

> **We** can also set up some of the preferences that you may have overlooked.
> **Our next step** will be to work with those who have not converted.
> **We** will contact you to discuss the concerns that you have. (Email 5)

We can see in this use of thematic instantiation a grammatical repositioning of OIT that asserts a new and more essential campus position. Situating his office's authority in thematic position supported the social change also initiated in the interplay of extrinsic modality. Working together, these two functions helped to constitute a key social change within Northeastern's social networks. These were changes that Martin had hoped the WorkWare implementation would achieve but he never overtly mentioned this desire nor did he ever make a semantic argument to augment his power on campus.

Throughout the implementation, the content (semantics) of Martin's discourse was ideological but not in the way, for example, that corporate control of universities and education is ideological. Like promoters of corporate education, Martin used intertextual references to ideological discourses (evangelical Christianity). However, he did not use these as ends, but as common discourses that emerged from his own subject position and from the social network that aggregated within his office and his personal community. It is important to note that his goal was not to 'convert' users to his religious beliefs. His use of evangelical discourse, like the use of religious metaphors to promote business process re-engineering, emerged from well-established discursive patterns that were dominant within Martin's own social network. Intertextual patterns, such as the use of religious discourse to describe technologies, are deeply embedded and though they are useful ways to help us understand both sources, they should not necessarily be confused or suspected to be promoting their originary discourse within their adopted forms. Their use will influence the adopted network but in contested and stochastic ways.

Martin's attempts to persuade people to adopt WorkWare emerged from his own personal experience and discursive resources. His beliefs provided him with a rationale, a vocabulary and a narrative with which to frame the implementation and to interpret his role in this transition. At the same time, his focus and ambition during the campaign were oddly similar to his faith-based assumptions and ambitions. As a 'believer' Martin understands and knows what the final results of this change are going to be. They are, after all, predestined. He has finite reasons for the change and these reasons are hierarchical. As such, Martin's discourse was not designed to destabilize a context, for him that has already occurred with the President's edict. Martin's challenge was to restabilize the context, to bring the world back into some sort of social order.

At the same time, Martin's discourse privileged specific social capital and a specific way of interpreting the social world. Yet, 'converting' to WorkWare did not necessarily mean that campus users adhered to a similar belief system or even acknowledged the discourse in which Martin participates. While users' actions may have reinforced Martin's belief system, their motivations for conversion or rejection could have nothing to do with Martin's evangelical discourse, their understanding or opinions about his technology, or his evangelism for his office. Thus, the 'faith' given over by users to the new software can be simultaneously ambivalent, situational and conditional suggesting that personal and public identity within organizations is a tenuous and brokered cultural capital. This capital is dependent on strategic alliance and a ready (yet agnostic) identification with stochastic and transitioning narratives and interests.

With WorkWare as with his personal faith, Martin need not understand why people would resist or reject his offering. For him, rejection means intentionally rejecting his community, being left behind as he and others who have accepted their election move on to their reward. Similarly, in his faith, Martin does not see his role as someone who will destabilize other belief systems so much as to lead and to provide an alternative for non-believers. Here, technology is neither the angel nor the demon, it is neither determining structure nor agent. Instead, it appears simply a terrain onto which long-standing beliefs and contests are again made manifest.

6 Linguistic politeness in technologized means

This project has been concerned with describing and better understanding the function and the implication of discourses that initiate, enable and stabilize processes of social change. Within this larger frame, I have introduced the concept of discourse technologies and their uses as a means for creating social changes. The previous two chapters examined sentence and clause-level features of texts that created an organization-wide social change and through this analysis I introduced several new text features that can be associated with technocratic discourses.

These features enabled change not through overt semantic persuasion, by developing a consensus within a community, or even through force of will, but by instantiating change discursively – in the aggregation and structuring of language, text, situation and materiality. Thus, the goal of introducing the micro-analysis of Chapters 4 and 5 was to connect micro-text features with ongoing macro or network-wide activities and to provide a functional description of how technologized texts work.[1]

Such an effort takes seriously the claim that texts construct and reflect social networks and social action. At the same time, this effort provides an empirical link and demonstration of how this process might occur. Working at the granular level and using features presented in the previous two chapters as a place to start, we can better account for how change occurs within social networks. Such studies can also differentiate and highlight the use of discourse technologies that enact social change. Such a project enables us to better understand how change is constituted (aggregated) within a social network.

Not all social change occurs exclusively through technocratic means, what I will call *textnologies*. However, social change requires persuasion and, since change involves the construction of 'the new', participants must at some point agree to what is not yet real. This aspect of persuasion, otherwise characterized as mystical, magical or faith-based, could also be considered technological in the sense that it is tool-driven. Here, rhetors use the tool of language to create what is not yet real. Rather than appeal to ethos, pathos or logos, the text constitutes the new as an already existing discursive phenomenon. In this way textnologies (or the technocratic moment) may be a necessary component of some processes of social change.

At the same time, textnologies are a common and routine part of modern organizational life. With the growth and popularity of hybrid management fields such as information technology and information science, technocratic approaches to implementing and managing social change within organizations and social networks neatly align management practices with the technological imperative taught in these disciplines. These fields, along with technology-based management practices such as supply-chain management and business process re-engineering, are constituted by the use of information technologies, statistical efficiencies and strict production, storage and shipping rationalizations (Case 1999). Thus, the direct application of the text analysis applied in the previous chapters can be the examination and critique of textnologies as employed by and within such management approaches and the examination of their spread and utility in other social networks (churches, nonprofits, social groups, schools).

A functional orientation for strategic discourses

In proposing that we examine the spread of textnologies beyond their application in purely technical or IT management networks, I want to point to ways this discourse has influenced a much wider field of social interaction. Although the semantic content of the WorkWare implementation dealt with technical issues, these discourses are not unique to technical or even solely functional work sites. The discursive forms we have seen used to initiate change can reorient 'textnologies' in more functional and performative ways. This point will be further elaborated in the concluding chapter, but for now I want to claim that perceiving technical discourse (technical communication) functionally rather than topically (texts about technology) has important implications for the various networks constituted by these technical discourses. A functional definition of technical discourse orients us better to recognize and examine the expansion of technocratic means in sites beyond immediately technical or scientific workspaces and it does not limit the study of technical discourse merely to texts-about-technology.

The process whereby technical discourses have moved into other social networks beyond text-about-technology has been described by Fairclough as the 'colonization of orders of discourse' (Fairclough 1992) and alternatively by Habermas as the 'colonization of the lifeworld' (Habermas 1987) by instrumental rationality. What these terms refer to is the discursive act whereby networks are constituted by technocratic discourse even though these networks might have little to do with actual material technologies. For example, as we have seen in the case of the capitalist marketization of academic sites, ideological advocates have employed technocratic discourses to standardize behaviours, routinize daily activities and change employee and client relationships without offering substantial reasons or arguments for these changes.[2]

If, as I have theorized, discursive actions reflect and constitute social and material practices, the use of technocratic discourses, even in non-technical spaces, would have consequences for social relations within these changed sites. As I noted in Chapter 3, the imposition of technocratic values has been perceived as a significant problem by humanities scholars and others concerned about the influence of technology and technical processes on everyday life. For Habermas, the extension of instrumental rationality into what he called 'the lifeworld' directly threatens those discursive processes and networks that sustain the life-world (Habermas 1987, pp. 280–1; see also Simpson 1995, pp. 84–5). And, indeed, Habermas' critique has been quite influential in other examinations of discourse technologies. As reviewed in Chapter 2, the criticisms of technologized discourse by Lemke, Fairclough and Katz clearly echoed Habermas' concerns about the rationalization of social relations. Recall, these writers argued that textnologies:

- elide issues of choice in decision making and deflect the social, political and moral implication of decisions (Lemke 1995);
- are designed, tested and deliberately structured to create the appearance of rational and standardized solutions to problems (Fairclough 1992); and
- promote expediency as the sole value of decision-making and action (Katz 1992).

And yet, as we have seen throughout the WorkWare study, textnologies are not autonomous. They occur within and emerge from social motivations, narratives and belief systems. Equally, technocratic rationality must occur as an aggregation of networks, not as an isolated and determining process. Technologies, like discourses, translate and aggregate even though in their successful application they make these very same processes invisible (Latour 1996, p. 76). How then can we characterize and explain the social relations and dynamics constructed by textnologies?

Do the social networks constituted and managed predominantly through textnologies have different or unique requirements and expectations for social relationships and behaviours? Or, as many writers in this area have argued, do textnologies necessarily colonize and thereby rationalize and dehumanize interpersonal and social relationships? If so, why do people allow such discourses to propagate and why are textnologies not more overtly resisted and questioned by otherwise knowledgeable audiences?

To pursue this question, I will turn to the linguistic notion of politeness as re-theorized by Richard Watts (2003). Although politeness theory may not seem to be an obvious place to initiate such an examination, Watts' critique of feature-based approaches to interpersonal behaviours and his location of (im)politeness within concepts of relevance and social currency (capital) articulate a rich way to examine not only the ways but also the possible reasons why textnologies constitute and reflect potentially preferred social relations within their social networks. For some, the notion of politeness may seem quaint and ill-suited to discussions of technology and social change. After all, (im)politeness is typically associated with superficial and stylistic performances, etiquette rather than persuasion or power. In what follows, however, I will be using the term (im)politeness to describe the larger scope and interpretation of social practice and persuasion and the interpersonal dynamic within persuasive contexts between and among rhetors and audiences.

For my purposes, the question of politeness refers to the social, interpersonal and relational restrictions that apply to the generation of discourse. My assumption is that persuasive discourses, discourses that enable social change, must, on some level, be recognized as 'polite', or at the very least not offensive, by the affected audience. Consequently, 'impolite' discourses theoretically would be rejected by audiences and therefore would not be able to achieve social change. This conceptualization of politeness understandably raises a problem for change practices that rely on technologized discourses since these discourses seem to enable social change but they do not necessarily admit or reveal the changes they evoke. In other words, they should be considered impolite or even offending discourses. Yet, in most cases, audiences rarely challenge textnologies.

Before getting to this work on linguistic (im)politeness, I will first summarize our working framework of technologized discourse by reviewing the discourse features examined in the WorkWare analysis in the previous chapters. I will then compare these features to a new change-management effort introduced by Martin three years after WorkWare (Figure 6.1, email 2006-01). This comparison will build a more detailed, continuous working model of discourse technologies within the social network at Northeastern and will point to the way these features continued to operate across the social network during subsequent activities of social change. From this discussion, I will transition to an examination of Watts' critique of politeness and his elaboration of a new approach to investigating and understanding politeness as enacted social practice. The chapter will then conclude by returning to this question about what constitutes appropriate and offensive socio-discursive conduct within technocratic regimes.

For reasons I will explain in the concluding chapter, I should emphasize here that this discussion does not intend to create a generalized theory of linguistic politeness nor a universal account of discourse technologies. The study is intentionally limited to the specific network at Northeastern and the specific social changes occurring during the WorkWare implementation. While my analysis may suggest various features, strategies or categories of interpretation that may be applicable to other studies and social networks, I wish to remain within this singularity and to limit my explanations and analysis to these specific forms and events. Thus, the category/form 'textnology' would be applicable to other social networks and contests of change. Similarly, the forms and negotiated behaviours I will introduce below for explaining 'polite' behaviour may resonate with other exchanges and discursive activities. However, I do not assume that the specific forms in which textnologies on the one hand or polite behaviours on the other appeared and were enacted in this study will necessarily be replicated in other social networks, spaces and times. They may indeed be replicated in other social networks but that is the task of other studies to document. Once we have developed a common language to explore the intersections of discourse and social change, our questions must then approach us as empirical problems that require empirical solutions.[3]

Discourse technologies and social change

The discourse technologies introduced and examined in the previous two chapters are summarized in Figure 6.1. The table works horizontally across the three forms of meaning constitution: textual (syntax), interpersonal (pragmatics) and ideational (semantics). In providing this summary, I am not suggesting that these represent an exhaustive list of discursive features associated with changing social networks. For example, other studies have highlighted the function of nominalization (Lemke 1995; Giltrow 1998; Fairclough 2005) and deixis (Ward 2004) as textual constituents that may also be associated with social change. Instead, this table offers a generative summary and starting point for future studies mapping textual features to activities promoting or resisting social change.

While we can isolate specific text features for the purpose of discussion and categorization, it is important to remember that discursive features always work in aggregate as part of social and discursive systems (Lemke 1995, p. 118). For example, Chapter 5 demonstrated a subtle technique of enacting social change through modality and the construction and resolution of strategic ambiguity. Recall that the strategic ambiguity embedded within modality was resolved by accepting an ideological presupposition. For example, in email 5 the presuppositions aggregated OIT's legitimacy and expertise, situating the office as an important actor on campus: 'We can also set up some of the preferences you may have overlooked' (email 5, sentence 7). Admitting the presupposition – the user (may have) overlooked WorkWare preferences – resolves the modal 'can' in a way that admits the technical expertise of Martin's staff and accords a degree of power to them based on this expert knowledge.[4]

Similarly, the sentence 'If you have specific questions, please call the helpdesk and we will have a technician work with you to explain the changes' (email 5, sentence 22) requires readers to admit that resistance or questions regarding WorkWare are solely technical issues which they do not understand. But these issues can be resolved through explanation (more information). Admitting the presupposition reinforces OIT's status as a knowledgeable and legitimate campus technology office. Materially, admitting the ideology (OIT is now a credible agent on campus) rewards the user with fewer hassles and, most importantly, the continuation of

textual	Clausal dynamics (synoptic/dynamic)	Copula verbs with subject complements	Given-new & theme/rheme juxtapositions (including determiner function)	Modality & progressive aspect (volition/extrinsic)	Resolution of strategic ambiguity (including ideological presupposition)
interpersonal		Identity construction (interpersonal relationships)		Intertextuality & discursive motivation	
ideational		WorkWare software installation	WorkWare server installation	Legitimacy of OIT (unstated)	Social network power for OIT (unstated)

Figure 6.1 Discursive features associated with social change during the WorkWare implementation

email service. Socially, accepting the ideology positions the user as part of a team and part of a renewed social network on campus.

Again, this is not to claim that this discursive process is so much trickery or deceit but my purpose is to stress that these features work together as technologies within and across specific material dynamics as they enable change. Similarly, we must be careful not to assume that specific discourse features necessarily construct social change. For example, we cannot argue that modality must always be associated with strategic ambiguity, a changing social network, or that all uses of modality indicate technologized discourse. In some cases, modality may be used to support the status quo or may have nothing at all to do with contexts of social change. In other cases, modality may be used to represent a genuine choice, dilemma or constraints on action.

The point is that, when used across certain material networks, we can empirically describe how various micro features structure conditions that lead to changed social networks. We can then see how technologized discourse enables social networks (e.g. the role of OIT on Northeastern Tech's campus) to change without overt discussion, input or acknowledgement that the social network should change or requires change.

Continuity and change: three years later

Three and a half years after the WorkWare implementation, Martin sent the following email to all computer users at Northeastern Tech.

From:	martin@northetech.edu
Subject:	Campus-wide Migration from Novell to Windows Active Directory - oit1134.06
Date:	23 May 2006 9:59:40 EDT
To:	campus_community@northetech.edu
Reply-To:	martin@northetech.edu

Summary: Northeastern Tech will be moving to Windows Active Directory for file storage and printing to replace the Novell system.

Impact: To prepare for this move, OIT staff will be visiting each computer on campus. We will be configuring network printers as necessary. We will also be recording your version of WorkWare and Symantec Antivirus, so that we can make necessary upgrades at a later date.

Timeline: We plan to begin making office visits on Tuesday 30 May. We will send out individual reminders as your scheduled visit approaches.

If you have any questions or concerns, please contact the OIT helpdesk at x-HELP or helpdesk@northetech.edu.

OIT Email Communication Number: oit1134.06

Figure 6.2 Email 2006-01

From clausal analysis, we can see that this is primarily a dynamic and changing situation. The email's sentence:clause ratio is 1:2.14 which is identical to the average ratio over all seven emails in the WorkWare campaign. The ratio is also higher than Vande Kopple's identified dynamic (1:1.82) data set from the *Physical Review*. Though Martin's text constitutes and reflects a dynamic context, he seems somewhat more confident now than when he initiated the WorkWare implementation. The 1:2.14 ratio of 2006-01 is lower than the initial introductory WorkWare email (1:2.4) and lower than the two emails (emails 4 and 5) which were written at the lowest point of the implementation (1:2.9; 1:2.4).

Working from the text features reviewed above, we can see that Martin's new change-text (2006-01) relies largely on the same system of discourse technologies to initiate and implement his changes. As we saw in the WorkWare implementation, modality instantiates much of the transitional activity. Specifically, the text relies on 'will' as a predictive verb often coupled with the copula (will be). The combination creates a predictive hedge that largely relies on OIT's authority for resolution.

Unlike the WorkWare texts, there are few presuppositions here. As such, the modals are situated as extrinsic, suggesting that users have little volition in this process. The most dynamic sentence in the email is sentence 4: 'We will also be recording your version of WorkWare and Symantec Antivirus so that we can make necessary upgrades at a later date'. The dynamic structure indicates greater volatility which is supported by the conditions of the social network. Sentence 4 is the only sentence that could involve some user volition, cooperation and potential resistance as it involves OIT's staff directly accessing the campus user's desktop. The sentence also includes a double modal 'we will' and 'we can' which, while not quite acknowledging user consent, suggests that some complicity would make the process much simpler. At the same time, the presupposition, 'necessary upgrades' embeds the utterance in relations of power and materiality – if users admit the necessity of the upgrades, they also admit OIT's authority, expertise and right to manipulate the user's personal computing environment. This admission enacts a strong extrinsic resolution to the two modals 'can' and 'will'.

The given and new patterns successfully achieve cohesion across the full email. Sentence 1 initiates with the name of the university (given) and places the technical change in the new sentence position. Sentence 2 then moves the technical change to the given position (formally in an adverbial clause) and codes it as an assumption 'to prepare' and a pronoun (this move). By sentence 3, the change is fully integrated into the technical vocabulary of the email (configuring). Sentence 3 also introduces a secondary change (recording your version of WorkWare ... make necessary upgrades at a later date) that has greater social overtones (surveillance, technical control over personal systems). Here, OIT's technical changes also result in changes (or reinforce already present conditions) within the social network and users' control over their workspaces and technologies.

The email does not include substantial intertextual references nor does it construct overt new interpersonal relationships. The promise to 'send out individual reminders as your scheduled visit approaches' seems to suggest some sort of ongoing relationship but there is nothing that semantically builds on this suggestion. Indeed, the promise of the reminder is asserted within the construction of an extrinsic action – a visit that approaches. Users have no choice about the visit nor do they appear to have any control or volition over when the visit will take place. Their guest will come over uninvited. 'Your scheduled visit' is also depersonalized as the event is given agency: the visit 'approaches'. Animating the visit this way, 'we will send out reminders as your scheduled visit approaches', also asserts control over the time and space associated with the technological project. OIT asserts an ability to condense, expand and control time as they plan the implementation of their new product (Simpson 1995; see discussion on technology and condensing time in Chapter 3).

OIT's empowered social position on the Northeastern Tech campus is more evident in this new email. Unlike the WorkWare campaign, Martin's claims are not situated within any other campus authority. In this case, he speaks on behalf of Northeastern Tech. He also is sufficiently empowered to proclaim the change to the new technology without any campus-wide discussion. He offers no semantic reasons for the move to Windows. Martin is also sufficiently enabled to use the term 'necessary' to justify his staff's actions without any further explanation, 'We will be configuring network printers as necessary' and 'we can make necessary upgrades'. Martin's text is functional, and fully technocratic – his discourse instantiates the change to Windows Active Directory and simultaneously reinforces his office's power to influence computing systems and users' own technologies across the campus.

Linguistic politeness and discourse technologies

In Chapter 4, I argued that the intentional manipulation of text features (for example deliberately manipulating the ratio of dynamic and synoptic clause structures to assert greater contextual stability) might violate established considerations of polite discourse and thereby make such approaches theoretically unacceptable. Using Grice's principles of politeness, specifically the 'principles of cooperation' (Grice 1975), I argued that we expect communicators to adhere to constraints of quality and content that mark obvious deviations or deletions as problematic or disingenuous. Even though Martin's discourse is fully technologized by email 2006-01 (reprinted above), his writing does not quite attempt the sort of consciously manipulative 'sleights of hand' I described in Chapter 4 as problematic. This does not mean that after reading my account he (and others) might attempt to reduce clausal dynamics in their organizational writing. But at the same time, we can ask whether such an approach would be strategically and situationally legitimate or successful.

In other words, our understanding and deployment of strategic discourses leads us to ask what sorts of social or relational restrictions apply (if any?) to discourse within changing networks. Do writers have a so-called 'clean slate' for discursive strategies? Or, do the specific aggregations that constitute social and organizational networks put various strategic restrictions in place that predetermine what would constitute an appropriate or legitimate argument? If so, how might such restrictions be enforced? Or, do specifically technical discourses successfully elide the sorts of interpersonal regulations (politeness conventions) we have traditionally established to govern interpersonal interactions? Have we abandoned Latour to return again to an ironic and even more problematic Habermas?[5]

I would like to initiate this discussion of linguistic politeness by building an initial framework for distinguishing various types of linguistic behaviours and by introducing a working vocabulary for examining this issue. Scollon (2001) used the term 'mediational means' to describe the cultural tools people use in social situations to achieve distinct purposes. Scollon cited as an example the act of purchasing a morning coffee which requires a sequence of practices: standing in line, making an order, waiting, standing in another queue with others also waiting for their drink, taking the coffee in turn, having verbal exchanges with others in the shop and then drinking the coffee either at a table, in conversation with others walking down the street, or alone (Scollon 2001, p. 111). To this list we could add various interpersonal activities such as asking someone if they are finished with a newspaper lying on a nearby table, asking for a cloth to clean up a spill, or answering (or choosing to ignore) a mobile telephone call while in the shop.

These events take place as 'a collection of rather specific and narrowly defined practices' (Scollon 2001, p. 111) and as Scollon put it, 'it is a blinding simplification to say that [he has] bought a cup of coffee' (Scollon 2001, p. 115). Each discrete event is governed by what we can call quite specific cultural tools used by each person within the coffee shop – the person taking the order, other customers, the person making the coffee and Scollon himself as a participant in his morning ritual. These tools are both material (the coins, room architecture, furniture, coffee, cups, time) and discursive (requests, clarifications, conversation, apologies). Using terminology developed by Wertsch (1998), Scollon used the terms 'cultural tools' and 'mediational means' interchangeably to identify the 'ways in which sociocultural and historical process enters into the actions of the person in society' (Scollon 2001, p. 119).[6] Ultimately, Scollon noted, these processes become routinized (internalized) within the person until they are accomplished innately, meaning without conscious interpretive effort.

As methods of successful social action, mediational means (or cultural tools) can be linked to what linguists have called 'politeness strategies' but in situational and distinct ways. As Scollon noted, there is a difference between 'social practice' and 'the social act' (Scollon 2001, p.119). Social practice identifies 'extended, historical processes'; in other words, time-honoured, situationally situated accepted methods of acting. Social acts identify 'irreversible, one-time-only, real-time action' (Scollon 2001, p. 119). In aggregate and over time, social acts become social practices, habitual personal practices naturalized over time. When generalized and objectified across a social network, social practices can be identified as cultural tools or mediational means – common ways in which agents have learned to successfully achieve specific actions within a social network. The goal of a theory of politeness then, in the terms used by Scollon, would be to articulate how, within a specific network, situational social acts (or aggregated social practices) function within or against extant and objectified mediational means (common understood ways of behaving).

Watts' theory of politeness as social practice

Generalizing from Scollon's model, we can position mediational means as a type of value. Let me explain. These linguistic behaviours enable social action, they help people to work with each other and to accomplish individual and group goals. Therefore, knowing the proper mediational means and how to best use them appropriately in a variety of social settings is a valuable social resource. The greater the scope and repertoire of one's mediational means the greater one's ability to operate in multiple social networks. This is the model Richard Watts (2003) outlined in his critical study of linguistic politeness. Except, for Watts mediational means are literally currency that people expend and exchange to achieve various social actions. From this model, which I will develop below, what we would call 'polite behaviour' is actuated as a form of payment, an exchange of capital from which people hope to derive a benefit. As Watts put it, politeness 'is *payment,* but it is payment *in excess* of what is ordinarily required by the politic behaviour of human action' (Watts 2003, p. 152, emphasis in original).

Watts argued that for too long the emphasis of theories of politeness has been on explaining why certain linguistic features, styles, vocabulary and semiotic behaviours should be regarded to be 'polite'. Instead, he proposed that linguists should explain how the participants within a particular network work together to create and position themselves and each other in polite and impolite ways within their specific networks. He argued that such research 'should aim to explain how all the interactants engaged in an ongoing verbal interaction negotiate the development of emergent networks and evaluate their own position and the positions of others within those networks' (Watts 2003, p. 255). Such an approach would place issues of politeness within a larger theory of situated social practices, moving it away from the characterization of individual behaviours and activities.

Historically, the predominant forms for studying linguistic politeness in English discourse emerged as an aggregation from concepts introduced by speech act theory in the 1960s, from the models of face and facework developed by communication scholar Erving Goffman and from Grice's work on conversational implicatures (Watts 2003, p. 53). Watts noted that, prior to this aggregation, several writers developed frameworks for understanding linguistic politeness in English as early as the German Romantic movement in the nineteenth century. This effort attempted to bridge the study of language with notions of the psychology of emotion and national character (Watts 2003, p. 54). Watt's history (which is itself a change narrative) seems to employ its own sense of retrospective continuity in the way he profiled what he calls *prepragmatic* approaches to politeness research. Without fully elaborating this history, which is available elsewhere (Watts 2003, pp. 53–6; Held 1992), it is sufficient to summarize Watts' point that these early studies of linguistic politeness attempted to situate language use within a specific social context. Initially, politeness was attributed to nationality and national character but later to more individual and situated social contexts.[7]

Watts connected these prepragmatic, situated studies of politeness to his own framework for understanding how certain behaviours function to operationalize social networks. In endorsing prepragmatic approaches to politeness, Watts built cases against the predominant rule and principle-governed forms for studying linguistic politeness as presented in Grice's cooperative principles (1975); Lakoff's (1973) extension of Grice's work to conversational maxims; Leech's (1983) systems of maxims and principles; and Brown and Levinson's (1978) work on 'face' and the 'face-saving' theory of politeness. Watts argued that these pragmatic examinations of politeness can be divided into two basic approaches, studies of the general conditions, conventions and social activities; and studies of specific forms of linguistic behaviour that are attributed to politeness. The first approach relies on the concept of facework, which roughly correlates with social activity type, while the second involves categorizing sets of expressions as inherently 'polite' (Brown and Levinson 1978, p. 119).

Both approaches are based with varying levels of dependence on Grice's cooperative principle which, Watts noted, is not surprising given the networks in which scholars in the field participated in the 1970s and 1980s. Grice's principle provided a systematic way to link speech act theory to linguistic pragmatics. Yet, Watts argued that despite the 'revolutionary nature' of the cooperative principle and its wide acceptance by scholars in many fields, the principle itself 'remains firmly anchored in the conduit metaphor of language in which language is a container with which a speaker can transfer meaning to a hearer' (Watts 2003, p. 205). In other words, the principles and their maxims describe barriers to meaning–making, assuming that conversation will be correctly interpreted so long as proper guiding maxims are mechanistically followed. The principles cannot overcome the influence context, situation, prior knowledge, experience or other variables may play in interpretation. In addition, the approach fails to recognize conversation as meaning–making, as a process in which participants actively construct and reinforce meaning as a social event. As Watts argued, 'if we follow the Gricean cooperative principle, there is no way of knowing not only *how* the addressee will derive the implicature, but even *what* implicature or implicatures s/he is likely to derive. We are, after all, *both* jointly responsible for maintaining the equilibrium of social relationships' (Watts 2003, p. 207, emphasis in original). The Gricean principles are thus not sufficiently dynamic to fully account for and address the stochastic and varying ways in which meaning is co-constructed in social conditions. In addition, Watts noted that the Gricean approach is also extremely culturally specific and many of the individual maxims do not hold in non-Western cultures.

In order to resolve the static nature of the Gricean cooperative principle, theorists used Goffman's (1982) notion of facework, suggesting that the ultimate purpose of the cooperative principle is to ensure that participants in a social network 'establish and maintain a mutual face' (Watts 2003, p. 203). Facework does provide a more dynamic model for modelling polite discourse as it attempts to account for the ways all participants in a conversation work to negotiate their own representations and the representations of others. Watts wrote that 'the negotiation of facework requires continual assessment of the lines expected in different types of social interaction and the kinds of face attributed to individuals by the interactants' (Watts 2003, p. 83). This activity requires considerable discernment with some situations requiring explicit politeness markers to ensure that members of the interaction are able to save or preserve 'face' (p. 83). And yet, even though in some situations the deliberate use of markers to manage face may preserve polite discourse, Watts noted that there is no actual correlation between rational, planned, instrumental behaviour and polite discourse. At times, people may use face-saving techniques but not for altruistic reasons or for reasons that have any concern at all for their audiences (Watts 2003, p. 82). Would such uses still be considered polite discourse?

In addition, some forms of facework (face-threatening, aggressive face) in specific contexts may actually function as polite behaviour. For example, I could say to a close friend, 'you are a dunce' which may not be interpreted as face-threatening. Similarly, Watts noted that a teacher may use an apparently polite (face-saving) expression to a student, 'would you mind taking your feet off my desk?' which actually reconstructs the social relationship in this situation to reaffirm the power of the instructor and the subordinate position of the student. Here, facework is not accomplished to induce politeness but to reinforce social power (Watts 2003, p. 114).

At the centre of Watts' critique is his claim that 'politeness theory can never be fully equated with face theory' (Watts 2003, p. 117). To this point, he makes the following assertions:

1. Linguistic politeness runs the risk of becoming vacuous since any linguistic expressions at all which are used in face-maintenance would be examples of linguistic politeness.
2. Certain speech activities which look and sound polite in fact aim to achieve the very opposite of face maintenance.
3. The definition rests on the assumption that all social interaction is geared toward cooperative behaviour (Watts 2003, pp. 118–19).

Thus, we can never be sure what we are witnessing or examining is actually politeness or some other communicative or social activity (Watts 2003, p. 119). As such, both predominant pragmatic approaches to politeness theory are inherently flawed. As emergent products of Grice's cooperative principle, the approaches assume static connections between specific behaviours and specific social intents.

Neither model has provided a sufficient way to empirically link social behaviour and intended meaning with specific actions, linguistic structures or symbolic activities. Secondly, both approaches are products of Western, European discursive systems. As such, neither approach can adequately address differing cultural norms or differing linguistic structures within unique social networks. This second critique can be expanded beyond Watts' geographical argument to also assert that neither approach is sufficiently dynamic to address polite behaviour as it may emerge within different sub-networks even within Western cultures. For example, different networks of work, sport, age or aesthetics may have developed unique methods of functional polite interaction. This leads us to theorize that social networks emerging from technological sites or technical functions may have aggregated different and unique forms of linguistic structure that may also function as polite interaction specific to that network.

In place of Gricean-based frameworks, Watts built what he called a 'social model of politeness' (Watts 2003, p. 142). Watts wrote that the purpose of such a model would be to explain 'how native speakers produce polite language' (Watts 2003, p. 142). Working from Bourdieu's (1991) concept of symbolic capital and the operations of the habitus, Watts argued that politeness, like money, is exchanged within social networks in order to achieve social functions (see also Werkhofer 1992). It is important to note that in building an explanatory model, Watts was not attempting to create a handbook or guide to universal politeness nor was he attempting to create a key for interpreting linguistic expression – a claim that some features are indicators of politeness rather than others. Instead, his approach attempted to explain why certain forms of discursive action achieve polite status and as such function as effective means to achieve social action. At the same time, his framework also sought to explain impolite discourse and its associate implications for social action. A politeness framework emerging from the study of social practice can demonstrate the utility and social action achieved by specific linguistic features (e.g. modality, presupposition) within certain social networks. But, a social practice framework will not assert a universal meaning for these features, for example arguing that a certain feature (passive voice) is (im)polite in all networks. Such a claim would fall back on the problems associated with Gricean politeness theory. Rather than focus on specific features then, we can focus on the use of symbolic capital within specific social networks.

Much like a prestige form of mediational means, symbolic capital refers to accumulations of prestige markers within a social network. Bourdieu wrote that 'the position of an agent in the social space can thus be defined by the position [s/]he occupies in the different fields, that is, in the distribution of the powers that are active in each of them' (Bourdieu 1991, p. 230). He listed these powers as economic capital, cultural capital, social capital and symbolic capital, which he also called prestige. These markers are not objective or universal but network-specific. This means that symbolic capital is based on recognition rather than overt production and it is a situation-specific recognition. All other forms of power, when recognized, are transformed into symbolic capital (Bourdieu 1991, p. 230).

Recognized forms of power (clothing, crown, guns) are 'objectified symbolic capital' and their effectiveness and durability are still subject to the specificity of a social network (Bourdieu 1991, p. 277). Bourdieu wrote that recognition of symbolic capital 'can function only if there is a convergence of social conditions' (Bourdieu 1991, p. 72) which dually assert and receive a recognition of importance. For example, Bourdieu wrote, 'only a hopeless soldier (or a "pure" linguist) could imagine that it would be possible to give his captain an order' (Bourdieu 1991, p. 75). A person may desire or believe that they have symbolic capital but the actual availability of this capital is determined and recognized by the social network. Similarly, the social network can delegitimize symbolic capital. This will happen when members of a social network recognize that a symbol is based on arbitrary function and then become aware of its arbitrary relation to the network (Bourdieu 1991, p. 277). Thus, Bourdieu asserted that 'the market' and not autonomous or independent features defines the appropriateness ('felicity') of language use. 'One must therefore assert,' he wrote, 'against all the forms of autonomization of a distinctly linguistic order, that all speech is produced for and through the market to which it owes its existence and its most specific properties' (Bourdieu 1991, p. 76).

Bourdieu's concept of power is similar to the post-feudal systems of power I reviewed in Chapter 2. Recall that as populations developed a consciousness of themselves as political, moral and civil beings apart from the sovereign, social power was no longer effective as an activity originating from a central point or being (a ruler). Instead, social power emerged as generative forces that produced methods of regulation invested in daily practice. Here, the dynamics of the market operate as one force among others that create and regulate conditions for appropriate discourse and behaviour. As self-conscious participants within market systems, people are capable of recognizing prestige linguistic and semiotic (discursive) forms. At the same time, these forms shift and change over time as their currency becomes appropriated by different social groups. Recall for example Labov's (1966) sociolinguistic studies of the socio-economic stratification of 'r' in New York. Here, Labov showed that the pronunciation of post-vocalic /r/ was stratified by economic status. As Foulkes and Docherty have stressed, this pronunciation feature and its association as an indicator of economic status is 'wholly the product of social construction', demonstrating that the connection between linguistic form and social category is 'arbitrary' (Foulkes and Docherty 2005, p. 3). To further complicate this relationship between form and its social activity, markers like /r/ shift over time. As more users recognize and adopt a specific feature or symbol, the value of that linguistic form declines and new features emerge to fulfil the prestige roles now abrogated from the former sign. Bourdieu wrote that this accommodation is not an act of overt domination or intimidation but a wilful but potentially unconscious disposition formed *over* time through experience, education and training within and across one's social networks. He wrote,

> The distinctiveness of symbolic domination lies precisely in the fact that it assumes, of those who submit to it, an attitude which challenges the usual dichotomy of freedom and constraint. The 'choices' of the habitus (for example using the 'received uvular "r"' instead of the rolled 'r' in the presence of legitimate speakers) are accomplished without consciousness or constraint, by virtue of the dispositions which, although they are unquestionably the product of social determinisms, are also constituted outside the spheres of consciousness and constraint.
>
> (Bourdieu 1991, p. 51)

People recognize and can relate the ongoing evaluation of the phoneme /r/ even though their own performance of the phoneme may differ (Bourdieu 1991, p. 52). This divergence in evaluation and performance suggests the situational regulation of social networks and the awareness that politeness may require performance of differing varieties in certain social conditions (for example, when in the presence of members of a different class so as not to assume membership in that class).

Bourdieu wrote that the 'definition of acceptability is found not in the situation but in the relationship between a market and a habitus, which is itself the product of the whole history of its relations with markets' (Bourdieu 1991, p. 81). He used the term 'habitus' to identify the 'set of *dispositions* which incline agents to act and react in certain ways (Thompson 1991, p. 12, emphasis in original). Much like Scollon's concept of mediational means, these dispositions result in regularized practices, approaches and interpretations that are considered normal or 'regular' but are not formed by any specific rule (sovereign). As the 'propensity to speak and to say determinate things ... which involves both the linguistic capacity to generate an infinite number of grammatically correct discourses and the social capacity to use this competence adequately in a determinate situation' (Bourdieu 1991, p. 37), the concept of habitus provides a direct link to a social production model of politeness. Habitus describes what is considered normal behaviour by which people accomplish routine tasks within specific social networks.

Watts presented Bourdieu's framework of habitus and social capital as a way to describe what he calls a theory of (im)politeness which, as I noted above, is an important aspect of a theory of practice (rather than a linguistic theory of symbolism). Watts argued that a practice-based model 'offers ways in which researchers can show when and perhaps why individual users of language in socio-communicative verbal interaction *classify* utterances as polite – even *express* utterances politely, and to allow both politeness and impoliteness to be evaluated by individual users' (Watts 2003, p. 160, emphasis in original). The difficulty with such a model, Watts acknowledged, is that 'there can be no objective criteria' for determining or explaining what can be considered polite or impolite behaviour other than past experience and the perceptions of individual respondents (Watts 2003, pp. 166–7). Yet, this does not mean that we cannot articulate functional reasons and practical explanations for why individuals consent to or resist certain discursive activities. Citing Bourdieu's theory of practice, Watts argued that 'practice is observable in instances of ongoing social interaction amongst individuals, which most often involves language' (Watts 2003, p. 148). But, he also noted that such interactions take place as the realization of 'objectified structures, products, modes of behavior which individuals gained through previous interaction (not always involving themselves)' (Watts 2003, p. 148). So long as we recognize the multiple ways (linguistic and non-linguistic) in which social practice is determined and enacted we should be able to account for politic behaviours.

Such an accounting would begin by situating linguistic politeness as the functional attuning of 'the demands of the market and the dispositions of the habitus' (Bourdieu 1991, p. 84). Turning to the examination of specific linguistic behaviours and features, Watts argued that linguistic behaviours (features) can be potentially interpreted as polite when they are performed 'in excess of politic behavior' (Watts 2003, p. 169). Here, Watts drew a distinction between politic and polite. By politic, he referred to the behaviours of the habitus (Watts 2003, p. 162) – regularized behaviours that attain normal and routine social goals. Discursive behaviour that does not meet the interpretive expectations of politic behaviour is considered impolite. Such activity does not conform to basic, regularized and accepted modes of expression and behaviour within a social network.

Polite behaviour then becomes, in his words, 'linguistic behavior that carries a value in an emergent network *in excess* of what is required by the politic behavior of the overall interaction' (Watts 2003, p. 162, emphasis added). Polite behaviour is behaviour that is valued more than the routine exchanges that constitute the habitus. 'No linguistic structures are inherently polite' Watts argued (Watts 2003, p. 168) but when measured against habitual actions in specific situations, certain features accord an excess of value which may then instantiate what we may call polite discourse.

What is considered (im)polite is thus a factor of the habitus of the specific social network, the individual(s) within that network and the linguistic capital these individuals are able to access, aggregate and deploy (Watts 2003, pp. 160–1). Watts recognized the considerable relativity within this model, and yet in order to address the multiple and conflicting empirical data that is interpreted as 'polite' such relativity and situational-specific modelling is required. For example, we know that power within a social network (such as an organization) is dynamic and changing. This means that what constitutes politic behaviour and by consequence polite behaviour must also be dynamic – what constitutes polite behaviour at one moment may constitute impolite behaviour at a different time and place. But, when combined with an understanding of the habitus, this model provides a way to use a fine-grained analysis of linguistic features to examine what constitutes both politic and polite discourse in a specific discursive and social network and situation and to determine simultaneously what and why specific forms of discourse are able to attain certain results within these specific sites.

Such an analysis also enables us to better code the operations of power within social networks since these operations are tied to the habitus and to actions precipitated in excess of the habitus. Power produces meanings and requires specific social practices. Given that such power dynamics are often latent, at times it may be what Watts called 'only our feel for what is acceptable in such a situation, i.e. how such situations are socially reproduced' that will enable us to interpret or perceive a specific behaviour or series of linguistic features as (in)appropriate (Watts 2003, p. 248). As Watts noted, 'we tend to notice the appropriate politic behavior only when it is violated' (Watts 2003, p. 248). Yet, as I will discuss in the concluding chapter, placed within an examination of what Rabinow called 'the anthropological contemporary' (2003) these tools provide us with the ability to make explicit and available our own habitus, or in other words, the habitus that governs and controls present day networks and power dynamics. As such, these examinations provide us with useful and functional insights into the power dynamics and formations that reside within and thereby construct our own social networks of exclusion, privilege and access.

Politeness, relevance and social action

How then do we apply this framework of politeness and social practice to our problem of technocratic discourses? Does such a relativistic theory prohibit us from developing and articulating ethics of social change? To answer these questions, I will return again to Watts' (2003) theory of politeness and to what he calls 'relevance theory' and its application for both politic and polite action.

A weakness in the social practice framework as presented to this point is that while the model describes the integration of discourse and social context (network) we have not yet described agency and the processes of interpretation at the level of individual audience members. Although the research tended to universalize audience experience and did not accurately explain audience activity, a strength of Gricean-based approaches was the emphasis investigators paid to audience interpretation and the resulting emphasis on audience experience as a defining characteristic of politeness. A model of politeness articulated as social practice requires an accurate theory of audience that can still be dynamic enough to operate within the larger framework of social action. To achieve this extension of his social practice theory, Watts re-articulated an approach described as 'relevance theory', initially developed by Sperber and Wilson (1995). Watts cautioned that his inclusion of relevance theory is supplemental to a larger framework of social practice, it is not a theoretical basis for a model of (im)politeness. As a concept, relevance allows us to explain the function of (im)polite discourse and semiotic action as social practice but it is a supplemental feature of the larger social practice framework (Watts 2003, p. 203).

As Watts explained, the inclusion of relevance within the framework of (im)politeness retains some interest in the Gricean (1975) maxim of relation. But, as was noted above in the criticisms of the Gricean model, Sperber and Wilson (1995) also expressed discontent with the model of cooperation and its resulting pragmatic emphasis on linguistic codes (Watts 2003, p. 208). For Sperber and Wilson, all of Grice's maxims can be reduced to a maxim of relation which states that an utterance should be relevant to other utterances that precede and follow. Or, as we saw in Chapter 3, Levinson's directive was to 'make your contribution relevant' (Levinson 1978, p. 102). Watts argued that this emphasis on relevance positions discourse as a stimulus which influences the knowledge within a social network. The discourse is then assumed to have at least one value which hearers (readers) must interpret for themselves. Watts argued that this point assumes that a discourse will always come with *'a guarantee of optimal relevance'* (Watts 2003, p. 209, emphasis in original), meaning that those producing the discourse will intend to alter their own and their audiences' knowledge. The extent to which the information alters this knowledge is up to the individual audience members. From this argument we can derive the point that relevance is determined by the audience and only in rare instances by the writer or speaker.

While an audience may determine certain discourses as relevant, Watts argued that no utterance 'can ever be fully determined with respect to its meaning' (Watts 2003, p. 210). Instead, the audience will bracket the propositional content of the discourse. This bracketed content will then be applied to the situation in which it was created and to the latent knowledge already known by the audience and the members of the network. This application results in meaning and in an interpretation of the relative value of the propositional content, the politic and the politeness. Thus, relevance is a sliding scale which can be interpreted differently by different members of the network (Watts 2003, p. 210). In addition, Watts noted that 'relevance may be negotiated from one conversational turn to the next' (Watts 2003, p. 210). Thus, participants in a social interaction are usually obligated to react to available discourses and from these reactions speakers and writers may interpret the relevance of their discourse. Watts argued that 'it is therefore up to the speaker to see to it that the other participants in an interaction expend as little effort as possible in inferring the most relevant information' (Watts 2003, p. 212). In an ideal discursive network, participants will exchange and revise a discourse until maximum relevance has been achieved.

However, as we have seen in the construction of change messages used at Northeastern Tech, these ideal assumptions of relevance theory may be slightly naive. The theory has no real way to explain why certain discourses would be intentionally scaffolded to appear irrelevant. I am thinking here of manufactured discourses like spin, intentionally misleading discourses, textual silences and certain textnologies that occur unnoticed. In these cases, there is no guarantee of relevance or more importantly not even any attempt to situate the discourse as relevant. Some discourses may even be constructed intentionally to appear irrelevant. Watts acknowledged that discourses may be devious. He noted a possible case when 'a speaker may be *paying* with politeness for devious reasons that give the hearer reason to doubt the sincerity of the utterance' (Watts 2003, p. 253, emphasis in original). Watts argued that in these cases most of us can recognize such deviousness as impolite.

And yet, we are still faced with the question of intentionally obscure discourse, discourse that has been intentionally technologized to achieve a specific social or material objective, and potentially crucial discourses that have been constructed to appear irrelevant so they are not noticed (a news release sent out the same night as a popular sporting match; public information intentionally released during the coverage of a high profile celebrity story; embedded ideological presuppositions designed to create the cognitive context for a social change). In some cases, we can obviously recognize and claim these practices as deceitful (even if we have too few discursive resources to effectively resist or describe such practices). Yet, in other cases, as we have seen throughout this book, people are less willing to challenge certain discourses, especially if those discourses enable us to attain certain social, material or technical goals (the effective provision of workplace email). Following the theory of politeness as social practice outlined above, it would seem that in these cases, people are willing to grant potentially impolite discourses polite status if the discourse instantiates a suitably relevant objective, that is, it accomplishes what the audience desires. To use Bourdieu's (1991) market-based terminology, people will exchange politeness if the resulting value is sufficiently relevant to exceed the level of potential offense. This is the other side of Watts' coin. Not only does politeness purchase social action, but as a symbolic commodity, we will forefeit politeness (or purchase impoliteness) if we value the resulting achievement. Thus, a technologized discourse may be impolite or even offensive, but if the social change it manifests is sufficiently relevant to an audience, they will be willing to accept the impolite discourse (or even grant it polite status).

This dynamic, however, is only available and relevant within a predetermined social and/or material context. As I noted in Chapter 3, these technologized practices are neither deterministic nor autonomous. They instead are aggregations of desires provided a means for operating. Before a social action can be fully technologized as a persuasive text, there must be a common interest within the social network for the realization of what that text will achieve. In this sense, the text does not so much cause the social change but enables it to occur. As Johnson-Eilola (2005) wrote, it 'participates' in the change. Martin could not successfully technologize Northeastern Tech's WorkWare implementation unless the social network on campus would be partially receptive to the new communications technology product. Receptivity could have many different motivations beyond the software itself – losing one's job, appeasing administrative hierarchies, indifference, attaining a job with Lyceum or a willingness to try something new. In addition, Martin could not have technologized an empowered social role for his office (OIT) unless there was already a desire (latent or pronounced) for campus leadership. The social changes that Martin created at Northeastern Tech were made available by needs already existing within the network. As such, the majority of campus users (including campus leadership) appeared to be willing to exchange the expectation and practice of polite discourse for control over computer technologies – which necessarily included control over the social context in which these technologies were used.

At the same time, the processes that enable social change will reinforce certain characteristics, behaviours and values across the social network in which they operate. Recall Latour's somewhat problematic assertion that a technological project 'gives itself a context' (Latour 1996, p. 133) by aggregating allies (Latour 1996, p. 71) and creating assemblages of metaphorical spokespersons (Latour 1996, p. 42). This process of contextualization will unavoidably manifest and reify the objectives of those who contribute. However, such objectives are not presented as individual and pure forms but rather as rearticulations, reconstructions resulting from their aggregation with other forms. In this way, individual self-interest becomes rearticulated in the technologized network as efficiency. Pragmatism is rearticulated into the separation of means and ends; the negation of aesthetics into instrumental means. In contexts of social change, those who are willing to exchange politeness for technologization do so because the product of their purchase is the realization of what they already desire.

The trade-off described above between politeness and goal actualization is evident in the following email written by Martin (2006-02).

> Subject: Student use of VOIP
> From: Martin <martin@northetech.edu>
> Date: 21-08-06 15:31
>
> As the academic and entertainment needs for bandwidth increase, our concern for the commodity Internet connection is magnified. With this in mind, we have been considering how to address VOIP traffic both in the dorms and in the computing labs. Since Northeastern offers its students the option to purchase in-room phone service and most students have cellphone service, we would like to prohibit VOIP traffic from the dorms and computing labs. The use of these services has been nominal over the last two years, but unless we implement this type of a policy, our expectation is that we would see marked increases in bandwidth utilization for these purposes. As always, we will deal with exceptions on a case by case basis.
>
> This issue was discussed last spring with students during the OIT forums and they understood the concern.
>
> Please reply with your thoughts by noon Wednesday (sooner if at all possible).
>
> Obviously we would like to be able to provide these services to our students; however, our current bandwidth connection does not afford us that opportunity without degrading our current functionality.
>
> We have spoken with the VP Student Services and the Dean of Students regarding this issue.
>
> Thank you,
>
> Martin

Figure 6.3 Email 2006-02

VOIP is an acronym for Voice Over Internet Protocol, a technology that allows users to make long-distance telephone calls over the Internet. VOIP technology bypasses local telephone services, meaning that users do not have to pay long-distance charges. The social and temporal situation for this email is important to scaffold before we more fully examine Martin's text. The email was designed for a different social network than the general announcements previously presented in this chapter (email 2006-01). This email about VOIP was written to a technology advisory committee at Northeastern Tech. This is the same committee that was responsible for overseeing Martin's implementation of WorkWare but now constituted by several different members. The committee is made up of rotating positions of 20 faculty, staff and administrative leaders from across campus who represent campus interests in technology decisions. The committee is basically a liaison between Lyceum and Northeastern. This social network is required to oversee OIT policy decisions and provide Martin and his staff with input and advice on information technology use across campus. The committee members could become key allies for Martin and his staff or they have the potential to disrupt his leadership and initiatives. The social network thus forces Martin into discursive actions that both inform and consult while still maintaining his central leadership function.

We can see the influence of the social network in the politeness indicators and the negotiated stance taken between informing and consulting. The email uses fewer technologized discursive forms than Martin's previous messages. Predictably, the email is most dynamic as he makes the case for not providing VOIP services on campus. The sentences throughout the first paragraph, where the case gets made, are heavily claused and contain multiple phrases. Martin appears to recognize the need here to create an argument and provide reasons for this policy decision. The argument relies on modality (would see marked increases in bandwidth utilization; would like to be able to provide these services) and it is embedded within several presuppositions (needs for bandwidth increase; students have telephone services available to them).

The email also follows a predictable given/new structure with the problem ('commodity Internet connection'; 'student use of VOIP') introduced in new structures and then transferred to given status ('these services'). The solution is also introduced as new information in dynamic style ('we would like to prohibit VOIP traffic from the dorms and computing labs'). From its status as new, the solution is turned into a policy in another new structure ('this type of a policy'). The policy is then transferred to given status in the second paragraph ('This issue'). The actual policy is not repeated in subsequent lines but is represented by elision ('with your thoughts') and pronouns ('these services', 'that opportunity', 'this issue').

Martin's discursive tension between advice and control is evident as he presents the solution to the problem. Whereas the email presents the problem in relatively overt discourse, the solution implementation is technologized through given/new structures. Rather than seek alternatives, approval or advice on how to deal with this issue, Martin immediately turns his solution (not providing VOIP) into a policy ('this type of policy'), a pronoun ('This issue') and then a textual silence ('with your thoughts'). Using the pronoun 'this' makes the solution ambiguous as it combines the problem (commodity Internet connection) with the policy (not providing VOIP). In asking for 'thoughts' on the issue Martin is not seeking alternative solutions and his penultimate paragraph is overt about the fact that the decision not to provide VOIP has already been made. To provide 'these services' would result in 'degrading our current functionality'.

Martin's dichotomy between VOIP and preserving the network's current functionality may be a false choice since the audience does not have sufficient information to judge his claim. As such, we could argue that he violates Gricean (1975) standards of politeness by not providing sufficient information (maxim of quantity). However, in technologizing this point, he rearticulates his argument into a claim of self-interest. Here, relevance does not refer to the relevance of the information presented (Gricean maxims) but the relevance of the information to the specific addressees within the social network. It is not in any member of the technology committee's self-interest to jeopardize their own access to bandwidth so that students can attain free long-distance telephone service. In addition, all members of the committee already have telephone service provided in their offices. Here, Martin's argument exchanges politeness for self-interest. Even though the lack of relevant information may be impolite, the argument to ban VOIP becomes acceptable to the members of this social network because it is relevant to their self-interest. The task of challenging Martin's claim about the limitations of current bandwith would be time-consuming and not in his audience's self-interest. Simply consenting to the policy is efficient and practical, two technical values that reflect and reaffirm the latent values within this social network.[8]

Students, supporting their own self-interest, could be expected to argue in favour of VOIP. However, in a student forum Martin again appealed to a different sort of self-interest by arguing that VOIP would 'degrade the network for academic needs' and that he was unsure if 'the current network can support some of the more network-intensive games'.[9] By technologizing the argument for implementation, Martin's discourse rearticulates the technical value efficiency into the social value self-interest. He is not forcing new values onto this community but rearticulating values that already currently inform and structure social relations within this specific social network.

Rhetorical studies have long been interested in examining and predicting what Aristotle called the available means of persuasion. These means of persuasion are usually defined topically – what the speaker is claiming, what the text reports, what claims are made and the relative effectiveness of these claims. Where rhetorical studies have been less effective has been in articulating the interplay between topical and structural functions of discourses that motivate human action.

This chapter has argued that motivation is often the result of both content and structure – that an understanding of what makes a particular argument acceptable entails understanding the form of the utterance as well as the content. In some cases, technologized arguments may be perceived to be offensive because they violate expectations of appropriate discourse. However, we cannot assume that all technologized texts are impolite or offensive, nor can we assume that simply because a discourse is offensive it will not be successful. Such discourses may be granted 'acceptable' or 'polite' status because they include relevant 'payoffs' and desires for their audiences. Relevance, as we saw in the case of the VOIP policy, refers to the way the claim becomes rearticulated within the social network.

Rearticulation of discourse points again to the conditions whereby discourse both reflects and constitutes social networks. Recall, a technological discourse that emphasizes efficiency and rationalization, the separation of means and ends and instrumentalism is rearticulated within this specific social network into self-interest, pragmatism and the negation of aesthetics/language, thereby translating and affirming desires already existing within the social network. Granted, this may not be an ideal or comforting explanation of the problem of successful objectionable discourses. But, as I will argue in the final chapter, providing grand narratives, universal critiques and hopeful solutions may not be the responsibility or even the possibility of a practice of situated discourse studies of change. There are times when we need to make the case that, even in the human sciences, there is value in (simple) explanation.

7 Conclusion: Change studies

> it was not sufficient to make a good soap,
> one had to train the people to wash
>
> (Schumpter (1964), citation in Rabinow 2003, p. 126)

This final chapter articulates a research orientation for studying the process, implications and politics of social change as discursive forms/events.[1] The chapter begins by examining and reviewing Classical concepts of change as evidenced by Niccolo Machiavelli's *The Prince*. Although Machiavelli is often considered one of the founding thinkers of modern politics, I contrast several key assumptions about his depiction of change and the more technocratic forms of change examined in previous chapters. Arguing that in Machiavelli we miss a 'technique' of change, the chapter suggests several ways the human sciences can engage the empirical study of social change, largely as a project of discursive study.

This project is necessary, I argue, because with the postmodern fragmentation of narratives, traditions and foundational beliefs, we do not have a systematic way of examining, critiquing or resisting non-narrative forms/events of change. The chapter concludes by reviewing four generative tensions that emerged from the WorkWare study in the previous chapters: the material and the discursive constitution of change; the constitution of discourse studies as multilayered, constructive and empirical; the study of social systems as stochastic networks (the study of singularities and the expectation of generalities); and the necessity of power and the limitations of ethics.

Concluding *Technology, Time and the Conversations of Modernity*, Lorenzo Simpson suggestively forecast a renewed nostalgia for the grand narrative as an organizing principle. 'Modernity' he wrote, 'marks the state at which totalizing worldviews rooted in religion and tradition lose their purchase on the European mind' (1995, p. 136). If modernity is to be demarcated by the fragmentation of master narratives and a nostalgic regret for the loss of continuity (of history, progress, time), it can also be demarcated by the conceptual emergence of change as an object (aim, purpose) of consciousness and eventually as a problem for human agency and intervention. With the fragmentation of what Foucault called 'temporal succession' (1972, p. 167), unifying tropes, narrative continuity and chronology, comes the possibility of change as a technology, as something specifically deployed to attain personal, political or organizational results.

It would not be accurate to claim that 'change' did not exist in Classical thought, as Foucault claimed about other modern concepts within the human sciences (Foucault 1970, p. 344). Instead, in Classical thought, change is constituted as an opening in narrative, a moment when the totality and continuity of narrative is temporally suspended. Change is a temporal and spatial pause that simultaneously presents opportunity and disaster. Classically defined, change is not a technique, a tool nor a technology. It is not an object nor a strategy. Instead, change is an opportunity/space. It is an objective space that is important for what it afforded or took away from power (the ruler).

Classical political discourse typically offers a reactive stance toward change, emphasizing continuity and adaptation in the space of change. The space of change is opened up by nature, fortune and chance. As Machiavelli put it, the adept ruler will anticipate change and should be prepared to respond to it. He wrote, 'Those princes who are utterly dependent on fortune come to grief when their fortune changes' (1961, p. 131). For, 'fortune is changeable' and the prince will prosper 'so long as fortune and policy are in accord' (1961, p. 133). Although the prince may not be able to predict or control when change may occur, the wise prince will use a time of peace industriously, so that when fortune does change, the prince 'will be found ready to resist adversity' (1961, p. 90).

There are perhaps degrees of nuance here. While acknowledging the consequences of nature, fortune and chance, Machiavelli's depiction of change was not wholly deterministic either. Thus, his view is important to a history of change because he contributed a necessity for human action as a response to partial breaks in the totality of predetermined narratives. In this way, I would like to position the perspective of change in *The Prince* as representative of Classical understandings of and approaches to change. In an argument that fittingly seems as much strategic and situated as radical, Machiavelli suggested that if fortune dictates a certain degree of experience, human activity can meet and influence that experience halfway. He wrote,

> I am not unaware that many have held and hold the opinion that events are controlled by fortune and by God in such a way that the prudence of man cannot modify them, indeed, that men have no influence whatsoever. Because of this, they would conclude that there is no point in sweating over things, but that one should submit to the rulings of chance. This opinion has been more widely held in our own times, because of the great changes and variations, beyond human imagining, which we have experienced and experience everyday. Sometimes, when thinking of this, I have myself inclined to this same opinion. Nonetheless, so as not to rule out our free will, I believe that it is probably true that fortune is the arbiter of half the things we do, leaving the other half or so to be controlled by ourselves.
>
> (1961, p. 130)

For some, this articulation of free will in the space of change signals a radical shift toward modernity as Machiavelli seemed to conceptualize a human agency that can augment fortune and the totalizing narratives of religion. But, we need to be careful not to impose our own notions of change and agency on Machiavelli. For Machiavelli, human action occurred as a response to fortune. Change may enable, or at the very least can elicit, human action but only as a situational response. As such, Machiavelli constituted change not as a human endeavour but as the temporal and situational conditions that elicit the enactment of power, typically through violence (conquest). He offered, 'in the antiquity and persistence of his rule memories of innovations and the reasons for them disappear; because one change always leaves a toothing-stone for the next' (1961, p. 34). As a temporal space, change enables and threatens power because it puts into play the potential for more instability – for further and ongoing disruptions of continuity.

Machiavelli respected the role and function of continuity and narrative. Note how his own statement of change, 'the other half or so to be controlled by ourselves', was deeply embedded within statements of continuity: 'This opinion has been more widely held in our own times, because of the great changes and variations, beyond human imagining, which we have experienced and experience everyday'; 'sometimes, when thinking of this, I have myself inclined to this same opinion'; 'fortune is the arbiter of half the things we do'. His contribution was to recognize that a pause in continuity is something that can both benefit and destroy the prince. He wrote, 'A man who is circumspect, when circumstances demand impetuous behavior, is unequal to the task, and so he comes to grief. If he changed his character according to the time and circumstances, then his fortune would not change' (1961, p. 132). In this case, the astute ruler recognizes the need to change behaviours to adapt to external forces.

But Machiavelli was less overt about how or why change occurs. He noted that space for change can be opened by malcontent and opposition. He argued that people 'willingly change their ruler, expecting to fare better' (1961, p. 34) and 'there always exist malcontents and those who want change' (1961, p. 46). Yet, we should remember, as Giddens has shown, 'there seems to have been no generic word for risk in Machiavelli's time' because concept of risk does not appear in European thought for another full century (Giddens 1991, p. 111). Therefore, 'Changing the ruler' is accomplished as an act of continuity in the temporal space opened by fortune. It is not a deliberate, strategic manipulation of the future. Nor is this a conceptual weighing of alternatives, estimating the potential benefits and risks of different courses of action. Fortune opens the possibility for a new ruler and 'malcontents' rush to fill the void in the expectation of better times.

What we see in *The Prince*, is not a fragmenting of narrative. Nor does *The Prince* signal the sort of crisis cited by Simpson. The prince's duty is to protect against the possibility of change, even in relations with personal staff and servants. Machiavelli argued that the effective ruler must ensure that servants, advisers, ministers and other personal staff are fully dependent on the prince. 'Having riches and honors to the point of surfeit [the personal staff] cannot but fear changes' (1961, p. 125). Although his references to Machiavelli are largely in passing, Foucault's assessment of Machiavelli's scope of control is worth repeating. Foucault wrote, 'Outside of God, outside of laws, outside of the great models given by nature, that is finally by God, outside of the principle of sovereignty, there is only the whim [caprice] of the Prince, there is only Machiavelli' (2004, p. 249; cited by Holden and Elden 2005, p. 3). The prince must enforce the very real experience that outside of the continuity that is the prince, there is nothing.

When fortune presents the opportunity for success and conquest, change must be readily ('impetuously') enacted. Machiavelli noted that under the proper circumstances, military action can install a new regime, creating a new political order. He wrote, 'I asked myself whether in present-day Italy the times were propitious to honour a new prince, and whether the circumstances existed here which would make it possible for a prudent and capable man to introduce a new order' (1961, p. 134). Change, a pause in continuity, presents opportunities and as such is something that must be met with immediate and swift action and force. It is something that must be 'commanded' (1961, p. 133). Yet, when this pause in continuity signals a threat to the sovereign, it must be subdued and quickly closed by an ongoing narrative of continuity, which, for Machiavelli, was the force and the rule (violence) of the sovereign. Recall, power, in the Classical sense, is deployed to sustain and retain its own sovereignty (Foucault 2003a, p. 164). In the same way that the Classical concept of power is constituted as a reactive use of force, Machiavelli could not assert change as a productive concept, nor could he conceptualize change as a technique which the prince may use as a benefit. At best, the ruler could anticipate that fortune may temporarily halt continuity, and the ruler must be ready to respond (typically with force) when that occurs.

Inasmuch as we miss in Machiavelli, in Foucault's terms, an 'art of government' (2004, p. 248; cited by Holden and Elden 2005, p. 3) so too do we miss an art of change, or perhaps more accurately, a *technique* of change. Machiavelli could articulate a technique of power but he was not able to conceptualize change outside of nature, fortune and chance. Although his depiction of change remains important for recognizing and asserting a need for human agency in the temporal space that fortune opens up, Machiavelli's change remains tied to fortune, and to continuity as a narrative function. At the same time, Machiavelli's conception of change remains very familiar to us. For example, Sullivan, an early nineteenth-century American educator writing 'to instruct the higher classes in schools in the origin, nature and use of political power', asserted that change is a natural act seen in the seasons and in what he calls 'animal life' (1831, p. 7). Sullivan deeply integrated and accounted for change as a normal and natural process. For him, change was continuity. Yet, in depicting change as continuity, Sullivan's depiction of change was not far from Machiavelli's in *The Prince*. Sullivan wrote, 'these recurring changes impose the necessity of being industrious, and of providing for wants which are sure to come' (1831, p. 8). From such actions arise 'the good order of society' (1831, p. 8). Sullivan's concern, interest and duty as an educator was to inscribe continuity, assurance and preservation. 'The power of preserving thoughts by writing or printing is another beneficent gift to man only' (1831, p. 10). Through writing, the 'treasures of the human mind descend through successive generations' (1831, p. 10).

Sullivan was not without the troubles of hope, expectation and uncertainty about the future. He noted that for him, 'The hope of a future life is natural to the human mind' (1831, p. 8). However, he was assured by his faith, 'the doubts and perplexities which hung over human hope, were at length dispelled by the revelation of Christianity' (1831, p. 8). From Christianity, Sullivan evoked permanence and assurance: 'From this source we are assured, not only that there are laws for human life, but also what they are; that they extend beyond this transitory scene; that welfare here, and future condition, depend on knowing what these laws are, and in rendering obedience to them' (1831, pp. 8–9). Thus, in this representation of change, change takes place to preserve continuity and to further the religious narrative which scaffolds Sullivan's experience and perception of daily life.

Studying change and the human sciences

Like the depictions of change from *The Prince* and Sullivan's educational treatise, considerable work in the human sciences (humanities, social sciences) has scaffolded social change with narratives of continuity. These narratives alternatively articulate religious, historical, cultural and philosophical continuities as ways to understand and respond to change. Such narratives present accounts that accommodate human experience to changing social, material and even natural events. Structured through these narratives, our conceptual understanding of change has remained relatively continuous since (and before) Machiavelli.

Simpson evoked this tradition when he argued that human action 'produces meaningful stories'. 'Like good literature,' these stories 'extend the range of scenarios in which our social, moral, and psychological concepts are at home, as they lead us to refine and rework the networks in which our concepts figure' (p. 65). Similarly, David Harvey has noted a preoccupation in the human sciences with the problem of continuity and change (1996, p. 96). He noted that in a tradition that extends from Marxism and various forms of psychoanalysis, through Foucault's later work on the body and pleasure, social theorists have struggled to locate the sources of disruption that might enable change to occur against the continuity of social structures. Harvey's analysis demonstrated the extent to which this tradition of social theory has taken continuity for granted. In this tradition, Harvey argued, ultimately and problematically continuity is interrupted and challenged by a 'residual' or a 'surplus' within moments of social process (1996, pp. 96–7).[2] According to this tradition, the residual is what challenges structural forces to enable change to occur, thereby enabling human agency. In Harvey's account, the residual is 'stubbornly resistant' to structure and is often, as in the case of Adorno's work (and many others'), associated with a sense of magic and myth (Harvey 1996, p. 97). Situated in magic, myth, poetry, the semiotic, the body or the experiential (see enumeration in Harvey 1996, pp. 97–8) the concept of the residual still promotes the narrative continuity of human experience. As it theoretically suggests a moment of indeterminancy that sparks and enables change, it reinforces permanence and continuity as the norm.

(Post)Modernity and change: crisis in narratives

Despite the key function of narrative as an interpretive act in the Classical dynamic between change and continuity, (post)modernity has placed the narrative form in crisis. As Simpson argued in the citation I used to open this chapter, narratives of continuity have lost their currency in Western rationality as a way to interpret human experience. (Post)modernity has declared the narrative form fragmented and has devalued continuity as nothing more than a myth of desire. This fragmentation of continuity is problematic for change studies because, in the Classical tradition, narrative and continuity have long been regarded as enabling conditions for change. If narrative and continuity have lost their value and function, if narrative no longer holds interpretive meaning and if continuity is simply a myth, what does this mean for our understanding of change? In large areas of (post)modern human endeavour, in organizational rationality, in sites of technology and technological exigence and in capitalist markets, narratives are no longer predominant or valued as interpretive devices. These networks, for which I will use the shorthand 'technocratic networks', have been constituted in ways that perceive little relevance or currency in narrative or continuity. Here, the Classical values of continuity and the residual have been displaced with alternative forms and processes for conceptualizing and achieving social change.

In technocratic networks, change is not a temporal space or a moment of opportunity within the continuity of narrative and it is not tied to notions of continuity, fortune or nature. Instead, change has become a technology and a tool, a means for achieving human agency. To use Rabinow's metaphor, technocratic networks have successfully 'purged the culture of its longstanding and unknowing dependence on, as well as its desire for, a world saturated with magical relationships' (2003, p. 110). Rabinow's use of 'magic' here calls to mind Harvey's 'residual', pointing to the moment of indeterminancy that leads to change. 'De-magification,'

Rabinow argued, denotes modernity's quest to excise narrative, story, and other 'mysterious forces at play in the world' (2003, p. 65). Demagification means that 'in principle, everything could be known and mastered by calculation' (Rabinow 2003, p. 65). Narrative becomes unnecessary and the residual becomes supplanted by calculus and technique. In (post)modernity, humans no longer respond to the (magical) space opened up by change. Instead, human agency invents, calculates and deploys change as a means for condensing time and controlling the future.

This concept of change as a technology is a problem for the human sciences as a research project because we have few ways to document, conceptualize or articulate emergent forms of change and their implications for social experience and community.[3] When examining and describing the WorkWare implementation, I have tried to resist the ready opportunity to demonize the technical-turn and thereby advocate for a return to a narrative-driven notion of change. As I followed the WorkWare campaign and came to better understand the effect of the change strategies on the college's social networks, I theorized that this seemed to be a different deployment of change which had not been fully documented by previous research. Accordingly, my motivation has been to provide an initial documentation and analysis of this emerging concept of change, including its discursive, strategic and ethical machinery. Following Rabinow, my suspicion is that newly emergent technocratic networks are not simply disdainful of the 'proliferation of totalizing systems' common in the humanities, but that, whatever their disciplinary legacies, most (post)modern discourses of management, technology and capital have been (re)constituted in opposition to the values and processes of the human sciences (Rabinow 2003, pp. 28–30).[4]

In this project I have proposed an analysis for approaching, interrogating and conceptualizing some forms of social change as technology, with the full meaning ascribed to that term in Chapter 3. As an analysis, my goal here is to provide an initial assessment for future projects and my bias throughout has been that new forms of change require empirical documentation and overt articulation before a political project can become, or be made, available. As has been a common problem for the human sciences, the deployment of change as a technology has outpaced our ability to effectively understand and address its implications for social life. In a helpful suggestion, Harvey cast this as a problematic of agency. For him, studies of change must move beyond articulating the residual, by which he means beyond denoting the position from which agency is enacted to incite change. Harvey proposed first examining 'how the idea of a "residual" or a "surplus" somehow outside of or beyond the hegemonic rules of determination gets set up' (1996, p. 104). In other words, studies of change need to examine *how agency is achieved* as a strategy and not simply what is argued from the position of agency. To accomplish such a study, Harvey suggested that 'it is not change per se that has to be explained, but the forces that hold down change and/or which give it a certain directionality' (1996, p. 105). Recognizing that change can emerge from any number of forces, a first problem for the researcher is to articulate how these forces initiate and sustain change. Harvey's argument is worth citing in full:

> There is no single moment within the social process devoid of the capacity for transformative activity – a new imaginary; a new discourse arising out of some peculiar hybrid of others; new rituals or institutional configurations; new modes of social relating; new material practices and bodily experiences; new political power relations arising out of their internalized contradictions. Each and every one of these moments is full to the brim with transformative potentialities.
>
> (1996, p. 105)

Given the numerous ways people might attain agency to spark change, it is important to articulate the different forces that people use to attain this agency. In the WorkWare study, I initially examined the institutional setting, networks and occurrences that existed before and during the WorkWare implementation on campus. From here, I examined the implementation itself through six interrelated dynamics that enabled the creation of a social field: discourse, power, social relations, material practices, institutions/rituals and beliefs/values/desires (from Harvey 1996, p. 78).

This assessment documented how agency was influenced and transformed by the integration of these dynamics across the temporal field of change. Thus, we saw Martin's agency constituted initially by claims to presidential authority and hierarchical ethos (institutions and rituals) in the early stages of the implementation. This approach was displaced by a second appeal to social relations and the organization of time as the product was rolled out. Then, as the software began to fail and users resisted the implementation, Martin appealed to users' beliefs and values, specifically to their identity as team members and members of the university community. Finally, Martin fully technologized the implementation both discursively and materially. By the final two emails, Martin had achieved sufficient agency to simply make the implementation and campus conformity a material act.

In these later stages, the implementation turned on the integration of discourse, power and materiality. This approach aggregated discourse as a technology and a tool of power. We can also more readily see in these final stages the complete fragmentation of narrative and the displacement of continuity akin to technocratic projects. The implementation discourse, both in the planning documents and in the implementation emails, rewrote the history of the integration (from software to server implementation) and created a retrospective continuity that articulated renewed objectives and rationale for the implementation (servers were getting old and needed to be replaced).

I should interrupt this account though to make an important comment.

It is easy and familiar to cast these events as a Classical change narrative. I have frequently had to stop myself from doing just that. For example, at this point in Martin's story, we could claim that the narrative of change failed and fragmented for both Martin and for a personified implementation campaign. The narratives of support within which Martin was operating had fragmented and dissolved leaving Martin isolated, threatening his social currency. Responding to the interruption in continuity, Martin, in a dramatic Machiavellian move, changed his approach and successfully retained his leadership and authority. Such an interpretation is powerful because it is so familiar. It Classically inscribes Martin's actions within a continuity of change scholarship, thereby reinforcing and extending this tradition. However, it is crucial to recognize that first such an account is empirically inaccurate. The WorkWare campaign did not initiate with Martin in a powerful social role from where he could sufficiently act or react according to the fortunes of change. According to some accounts, Martin was handed WorkWare and told to implement it. In addition, there was no continuity which Martin was tasked to extend. Second, and more crucially for the theoretical study of change undertaken here, interpreting the WorkWare implementation as a Classical narrative misses the *techniques of change* deployed across this social act. Enforcing a narrative perspective on these events would obscure Martin's strategic use of discourse and his proactive use of change as a social technology.

Rather than interpreting the temporal stages of the implementation as a narrative we instead can recognize the way Martin *used narrative* as a form (device) to create change. Pulling back the frame to the larger institutional network into which WorkWare was inscribed, I argued that for Martin and his office, WorkWare was not about the effective deployment of new software. Instead, it was about gaining a more influential and powerful social position on campus. WorkWare was a change strategically orchestrated to gain this social status. This is not to say that Martin planned to implement vapourware or that he expected to lose institutional support for his programme. Rather, these events were incidental to the larger goals the change was designed to achieve and these events were reflective of the stochastic network in which he was working (more below). As I mentioned previously, discursively, this project was about Martin gaining sufficient social power on campus to move from embedding his plans within a presidential decree (as he did in email 1) to simply stating what he intended to do (as he does late in the campaign and in subsequent years). Materially, the campaign was about producing and extending power from simply maintaining campus computer services to managing important academic and social issues on campus (intellectual property, access, appropriate uses of technology, firewalls, available technical services).

Examining the implementation as a problematic of agency enabled us to determine how agency was constituted by different and accumulating forces across the campaign. However, what it does not show us, and what we did not yet know, was how these forces were then articulated to achieve specific transformations within the social network. In other words, we can document various forms of agency during a change but we still need to fully articulate the techniques by which change was accomplished – how change was used to initiate, introduce and ultimately stabilize social action. In some ways, this means getting inside the 'black box' Bazerman noted is common with representations and speech acts (1999, p. 345). Thus, our second problem for studying social change becomes opening up the metaphorical black box and explaining how the technology works.

As a partial response to this second problem, I have attempted to provide a focused, granular form of analysis that examines specific discursive features and their functioning as techniques of change within the social network. Such an analysis seeks to apply a certain amount of analytical pressure on several key tensions operating within movements of social change. By tensions, I do not mean overt contradictions so much as complexities within and across social networks. While there are probably other ways to scaffold studies of social change, my study of WorkWare led me to four specific tensions that seemed to be useful for understanding the operation of change in this project. These tensions were: the material and the discursive constitution of change; the constitution of discourse multilayered, constructive and empirical; the study of social systems as stochastic networks (the study of singularities and the expectation of generalities); and the necessity of power and the limitations of ethics. In what follows, I will conclude this book by elaborating on these tensions and their implications for studies of social change.

Material and discursive constitution of change: textnologies

In his study of Edison, Charles Bazerman concluded that 'during the development of incandescent lighting Thomas Edison and his associates were busy people. They of course were making electricity and light. They too were making meanings. They were creating value for their endeavor' (Bazerman 1999, p. 333). Bazerman's point was that discourse and materiality are never far apart and in this book I have attempted to describe a specific form of discourse that has material consequences in communities and networks. In the previous chapter, I called these discourses 'textnologies'. Without subordinating one to the other, language scholarship has positioned discourse and materiality as mutually constituting social experience (Chouliaraki and Fairclough 1999, pp. 28–9). As we saw in Harvey's work (1996, p. 78), discourse is 'internalized' within other social moments (institutions/rituals, beliefs/values/desires, power, social relations and material practices). The problem here, as I have tried to address in this book, has been the continued sense, as Harvey put it, that language practices are 'about' and 'representative' rather than enactive and productive. I do not wish to reassert the 'discourse idealist' position (Chouliaraki and Fairclough 1999, p. 28; Latour 2003, pp. 29–30) situated within more extreme social constructivist approaches, which argues that the social (and in extreme cases the material) is nothing but discourse. Instead, I want to point to active discourses which operate as material agents and technologies and have actual consequences on social networks. Textnologies are functional, technical discourses, sometimes operating independently to achieve specific and strategic tasks. For example, a Web crawler is a technology that aggregates text as data, information and meaning by compiling discourses across the World Wide Web. These are texts that create their own texts, document their use, revise and update and provide information, value and meaning. As such, these textnologies are not reproductive or representational but creative and productive.

The discourses that constituted the WorkWare implementation were similarly productive. If we examined the WorkWare implementation as a series of representative texts we would only account for the persuasiveness of the implementation judged by how accurately the texts described the software, articulated user needs, and brought these together in a persuasive text. But, we would have no way, for example, to seriously account for why Martin sent email messages to the entire campus community about users who had not yet switched. Martin's office could easily tell who had already adopted WorkWare by simply looking at the servers' log files. Cross-referencing these against a list of campus users would reveal who had not yet switched. Thus, examining these emails as persuasive documents written to convince users to switch to WorkWare seems to miss much of the point. As Chapter 5 demonstrated, these emails operated as tools to create a community, to reassure users that OIT had the situation under control, and to establish and extend the legitimacy and authority of OIT across campus. By appearing to be dealing with holdouts, Martin's office gained considerable confidence and legitimacy from those who had already switched. In addition, these emails would also encourage users who had adopted the new software to stay the course and not abandon the project. In many ways, the audience for these messages was not the people who had not yet converted, it was those who were already part of the community.

Apart from their representative functions, textnologies are strategically deployed as tools with actual functions. These are technical discourses, enacting a different sort of persuasion, based on the structure and design of the message itself. To echo Derrida, this writing 'is not the means of transference of meaning, the exchange of intentions and meanings, discourse and the "communication of consciousness"' (1988, p. 20). In these cases, discourse and materiality appear to collide into texts that physically act in the world rather than represent or even construct the world. These are texts (if texts is the appropriate word) 'of which the system of speech, consciousness, meaning, presence, truth, etc., [is] only an effect' (Derrida 1988, p. 20).

Textnologies orient change studies toward theories of activity

(Engeström 1990; Spinuzzi 2003) but as ways to interrogate, examine and denaturalize the function and the nuances of activity. In other words, it is not sufficient for change studies to only claim that texts are activities. That much can be taken for granted by now. Instead, change studies must be able to empirically detail how textnologies take place and achieve events/ results. To reveal my own textnology here, I should admit my own critical project embedded within this claim. My suspicion is that textnologies function much like the magic cited by Harvey and Rabinow. Textnologies only work when people do not understand them or do not notice their operation. Because a textnology does not state its intention, its ultimate purpose and reason remains hidden (naturalized). However, once a textnology has been explained and detailed as a specific strategy for social action, it loses its sense of mystery and its power. As Simpson wrote about other genres of persuasion, audiences often judge arguments not only on the reasons for the proposal but also on the explanation given for its necessity (1995, p. 106). By definition, textnologies elide both reasons and explanation. Yet, once we denaturalize and question the purposes and the intent of a textnology, it can be challenged, resisted or even colonized. In addition we can then hold accountable the rhetors who deployed the textnology. Simpson put it this way, 'our assumptions about the purposes or intent of an expression will guide our selection of what is to count as a good reason for it' (1995, p. 106). In order to determine if there are good reasons for change, we must first denaturalize the purposes and intents embedded within the change.

This, then, can become the role of discourse studies within critical studies of social change. As I have attempted to demonstrate in Chapters 4 and 5, a detailed linguistic critique of change provides important analytical insight into the role and function and the purpose and effect of texts as they are used to create social change. Of course, not all texts operate as textnologies; language is a rich enough resource to provide multiple approaches to change. The challenge of a critical discourse-based approach to studies of change is to articulate, critique and possibly challenge whatever resources are deployed to create change, to describe how these resources function and to determine what purposes and intents their related changes are designed to achieve.

Discourse: multilayered, constructive, empirical

This project has attempted to connect critical discourse studies with critical studies of technology and sociological studies of change in order to extend and inform an understanding and awareness of the structural processes language can inscribe and enforce in contexts of change. A guiding concept throughout this project has been Chouliaraki and Fairclough's claim that the social is 'built into the grammatical tissue of language' (1999, p. 140). Following this claim, I have attempted to show how social moments (e.g. stability, instability, given/new) are reflected in and constituted by strategic discourse. Chapter 4, for example, examined dynamic and synoptic orientations throughout the WorkWare campaign, showing that the context of implementation was constituted and reflected by the grammatical cohesion patterns of the implementation discourse. Thus, language (discourse) at the micro-level must be a key feature of change studies. Similarly, without over-extending this concept, change studies must recognize discourses as constructive of social experience.

Paul Chilton has argued that a theory of language and politics (which is not far from a theory of language and change) must study cooperation and conflict as well as, in his words, 'how do the manifestations of power, language, conflict, cooperation come into being?' (2004, p. 198). What Chilton was arguing for here is a study of what he calls 'legitimization' (2004, p. 199). He claimed that political discourse is an attempt to 'get others to share a common view' on a wide variety of moral, economic and social issues. The way one accomplishes this, Chilton wrote, is linguistically. Chilton does not mean solely through words, but he is referring to actual linguistic structures. He argued, 'Probably all languages have grammaticalized modal expressions attached to concepts such as: social obligation–compulsion, certainty–doubt, evidence with credible–incredible sources' (2004, p. 202). In other words, motivating discourses are constituted within the structural units of the language. Thus, 'establishing "credibility", claiming "rightness" and "legitimizing" truth claims' constitute 'a political strategy that recruits many available linguistic mechanisms' (2004, p. 202). Chilton's linguistic study of political power shares much with a study of social change. Both projects recognize the central role discourses play, at multiple levels of constitution, in achieving social goals.

Fairclough has made a similar argument, specifically about studying social change within organizations (2005). He wrote that 'studies of organization need to include analysis of discourse' and he described what he calls a 'critical realist ontology' as a way to balance the dynamic between constructivist and objective approaches to such a project (2005, p. 915). A primary concern for his examination of the use of discourse analysis in organization studies is a tendency for organization studies to isolate and examine actions of organizing and organization as alternatively discursive and non-discursive projects. Fairclough argued that this is a false choice because both projects (organizing processes and organizational structures) are co-constituted linguistically. Thus, he argued that 'the relationship between discursive and non-discursive elements of the social should not be confused with the relationship between social processes (and isolable social events) and social (practices and) structures' because both entities are equally (yet not solely) constituted by discourse (2005, p. 918).

Therefore, when studying social change, we need to observe and examine the operations of discourse throughout the full moments of change. In addition, we must be attentive to the interplay of discourses, problematizations and materialities that create the social network and temporal situation. Although change can be diagnosed and traced discursively, Bazerman provided an important reminder that discourse cannot on its own constitute social networks. Recall that social networks are constituted by discourse and materiality. Change cannot be evidenced solely by discourse, but to extend and elaborate on Bazerman's claim, it must be evidenced by 'consequent actions' (1999, p. 341). What I mean here is that change cannot be fully evidenced as a cognitive event or a 'meeting of the minds' as rhetoric would put it. Change must be noted by an influence on the actions of people within relevant discourse networks (Bazerman 1999, p. 341).

Situating the recognition of change in consequent actions holds the study of change back from an orientation within an extreme form of constructivism. In order to be successful, the activities of discourse (textnology, narrative, persuasion) must be manifested by material actions within the social network. Latour has argued that this caution is crucial for the accuracy and legitimacy for the human sciences and I argue that it is especially important for a study of change. Initially, Latour noted that there is utility and import in a certain acknowledgment and understanding of constructivism as a process. 'Constructivism might be our only defense against fundamentalism' he wrote (2003, p. 28), pointing to a crucial hermeneutic and political project for modern scholarship. Yet, this perspective can be taken too far. 'The problem with construction,' Latour continued, 'is that no one could account for the building of anything, even the simplest shack, by using this metaphor as it has been popularized in social sciences' (2003, p. 31). He argued that nothing in the metaphor 'works':

> ... neither the role given to the builder or the maker, nor the role of the material used; neither the solidity and durability of the result, nor its contingencies or necessity; neither its history nor its lack thereof. If any mason, architect, any Little Pig were trying to build anything with the *theory of action* implied by constructivism, they would fail hopelessly to assemble any durable whole.
>
> (2003, p. 31)

Latour argued that as a social theory, constructivism needs to better articulate and account for the 'maker', the one who physically creates, invents, and builds; and who is aided and constrained by colleagues, plans, existing systems, prices of goods, quality of the software, education of the users, all of the elements that aggregate within the network.

Second, constructivism needs to better articulate a theory of 'making' by which Latour meant the dynamic faced by the maker who deals with 'the unexpected qualities and virtualities of materials' (2003, p. 31). Finally, Latour argued that constructivism has yet to properly account for the materials that are used to construct. 'Everywhere, building, creating, constructing, laboring means *to learn how to become sensitive* to the contrary requirements, to the exigencies, to the pressures of conflicting agencies where none of them is really in command' (2003, p. 33, emphasis in original). Thus, a theory of constructivism must re-address the issues of agency and materiality before it can assert a workable model for describing social events. To argue that social change is (partially) constructed by discursive events means we must understand and articulate the ways discourses interact and provide opportunities and resources for the people, material and specific networks that we are studying.

Fairclough's 'critical realist approach' to studying discourse provides one way to orient scholarship to better account for this sort of renewed focus on agency (2005). For Fairclough, discourse is constitutive; he wrote that he 'views discourse as an element of social processes and social events, and also an element of relatively durable social practices' (2005, p. 924). Though, as I have emphasized above, these practices and sites are not 'reducible to discourse' (2005, p. 924). Fairclough thus integrates Harvey's (1996) position that social relations, material elements and discourse internalize each other. The notion of 'critical realism' that Fairclough articulated is similar to Latour's perspective on constructivism. Critical realism asserts that the natural world exists apart from and is not dependent on human action. The social world is preconstructed for individuals; however, its construction does not mean that there are aspects of the social world about which we have no knowledge or understanding. Thus, as Fairclough wrote, 'we must avoid the epistemic fallacy of confusing the nature of our reality with our knowledge of reality' (2005, p. 922). Perceptions, however powerful, may not be accurate representations of events or structures and must be examined and understood within and against other kinds of knowledge and understanding.

The implications of this position are significant for the human sciences because the position asserts that not all representations of the world are equally accurate or complete. This means that we do not endorse a sort of experiential relativism that assures equal truth to everyone's experience. Instead, the framework aims at explanation, 'at explaining social processes and events in terms of the causal powers of both structures and human agency and the contingency of their effects' (2005, p. 923). Fairclough argued that critical realism differentiates among 'the real, the actual, and the empirical' as ways of being, understanding and knowing the social world (2005, p. 922). The real is associated with structures and 'causal powers', the actual concerns 'events and processes', and the empirical identifies that which is experienced by social actors (2005, p. 922). As these different layers indicate, advocates of critical realism differentiate between ontology (being, existence, physical objects) and epistemology (knowledge, belief). The relationship among these different forms of knowing is then mediated by social practices, including discourse.

Research within the framework abstracts from events in social life to the prestructured features of the social and then returns back to explain daily events, actions and processes. For example, we can take the regular daily event of an email communication (or a software implementation) and abstract from that process structural elements of the social world (language structures, religious beliefs and motivations, identity formation) and then return back to the specific social event to explain how and why the email functions as it does. Focusing on specific discourse features, a critical realist perspective can examine textnologies of change as mediated by the prestructured discourse features which the texts instantiate (intertextuality, nominalizations, given and new cohesive patterns, dynamic and synoptic clauses) and then return to the texts themselves to explain how and why the social change was enacted. Such a position allows for an examination of agency while still recognizing the constructive role pre-existing structures play within human experience.

Fairclough argued that by providing an explanation of social processes, critical realism provides change studies with a way to 'tease apart' different aspects and functions of change. For example, he listed new experiences, changes in discourse, changes in beliefs, changes in habits, and changes in organization as initial ways change can be differentiated. Each of these types of change can manifest uniquely and mediate different types of knowledge (ontology, epistemology). At the same time, given the iterative process of critical realism, we can describe the aggregation of forms of change when re-examining daily practices. As Fairclough wrote, and as the WorkWare study has shown, changes in discourse 'can and do contribute to change in beliefs, habits of action, and indeed organizations' (2005, p. 930). However, 'it is only by consistently regarding the difference between such elements that one can investigate the relations between them' (p. 930).

From Fairclough's account we can also emphasize the primary descriptive and analytical orientation of change studies. Unless brought to a problem specifically to provide an intervention (e.g. participant inquiry, action research; see, e.g., Faber 2002a), the student of change must be careful not to put forth a 'means of rectification' (Rabinow 2003, p. 18) but instead must carefully detail the constraints and possibilities at work within any given situation. As Rabinow put it, we must remember that the 'primary task of the analyst is not to proceed directly toward intervention and repair of the situation's discordancy, but rather to understand and to put forth a diagnosis of what makes these responses simultaneously possible' (2003, p. 18). At one level, change scholarship must be careful not to impose a solution on a changing network but to understand how a solution came to be – how it appeared, achieved legitimacy (delegitimized other solutions), became implemented and how, over time, it was sustained. This does not mean that subsequent studies, or applied projects would not employ certain staged studies to test various findings or theories of change within contexts of implementation. However, before such projects are possible, we require a fuller understanding of change(s) and its structure, function and mediation.

Networks and stochastic systems: the study of singularities and the expectation of generalities

Studying change requires a different focus for the human sciences. As I argued above, considerable scholarship in the human sciences has emphasized continuity, cohesion and integration. This emphasis has been especially apparent in linguistics where concepts related to discourse processes, systemic linguistics and even grammar itself orient analysis toward cohesion, connectivity and various methods for pattern recognition and text formation. Much of what we recognize today as discourse analysis, pragmatics and text linguistics emerged from and continues to build on foundational studies of cohesion (Halliday and Hasan 1976; Brown and Yule 1983, pp. 191–222). This work has been extremely influential for identifying structures, features and strategies that link units of discourse into cohesive units. As a result, most discursive features, including those I have used throughout this book, are predicated on the study of cohesion (e.g. presupposition, given/new, intertextuality, interdiscursivity, deixis, anaphoric/endophoric/exophoric references, field and orders of discourse). Even the concepts of frame, politeness and textual stability emerge from an emphasis on cohesion and continuity of discourse.

My suspicion throughout the WorkWare study was that a project examining the instantiation of a new technology is often a study of change. Thus, Latour could be writing about change studies when he claimed that the study of technological projects requires a 'move from a classical sociology – which has fixed frames of reference – to a relativistic sociology – which has fluctuating referents' (1996, p. 169). Similarly, Rabinow has called for a 'perpetually changing' new science in which 'new problems are pursued by new methods and truths are thereby discovered which open up significant new points of view' (2003, p. 36). Both claims for this new science find their own prompting in Foucault's discussion on change and transformation in the *Archaeology of Knowledge.* Here Foucault argued in favour of a new method that would try 'to establish the system of transformations that constitute "change"' (1972, p. 173). Foucault's *Archaeology* and its study of transformations provides us with an important way to emphasize the discontinuity of systems and the constitution of what comes to be called 'change' through the study of 'rules of formation'. The system of transformation includes rules by which statements can be formed, the creation of succession, and the conditions by which a correlation can exist between new statements and external events (1972, pp. 167–8). Thus, as Foucault's approach demonstrated discontinuity and showed how continuity becomes constituted in textual and material features, it created the potential for studies that empirically examine and explain random, stochastic networks. Archaeology 'is much more willing than the history of ideas to speak of discontinuities, ruptures, gaps, entirely new forms of positivity, and of sudden redistributions' (p. 169). But, to fully address issues of change, archaeology needs be supplemented with a critical realist examination of stochastic networks and their function throughout processes of change.

This is not to argue that we should abandon the concept of cohesion or that we should no longer demonstrate how social networks come to be seen as cohesive. Discourse studies has provided powerful ways of documenting the systemic construction of cohesion from disparate forces. But what we do not yet have are empirical forms for examining stochastic social networks – forms for examining social networks emerging from chance and randomness – that do not reinforce a sort of retrospective continuity onto the social formation. The concept of stochastic system is similar to the notion of Datacloud employed by Johnson-Eilola, who cast invention within similarly random associations of articulations (2005). Johnson-Eilola described the stochastic system this way:

> The sheer amount of information in our current lives – the infosphere that we individually inhabit – threatens our ability to sort it into neat categories, to summarize it in simple terms. Those necessities are falling by the wayside. Today, many of us thrive on information overload: e-mail, cell phone, satellite TV, instant messenger (IM). We live in a cloud of data, the *datacloud* – a shifting and only slightly contingently structured information space. In that space, we work with information, rearranging, filtering, breaking down, and combining. We are not looking for simplicity, but interesting juxtapositions and commentaries. This is the vague shape and erratic trajectory of the coming revolution.
>
> (2005, p. 4, emphasis in original)

Thus, Johnson-Eilola documented invention processes ranging from DJs digging through album archives, to the influence of IM, MOO, and other real-time interruptions on student projects, to spatial, architectural relations within physical offices.

Several current models for studying stochastic systems are available in physics, mathematics, biology and other interdisciplinary natural sciences. For the most part, these systems attempt to describe and model random motions of particles within and on various substances and dynamic (versus deterministic) biological processes. Other stochastic studies have examined dynamic and fluctuating events like stock prices, retirement securities (social security) and currency markets. Some statistical work has also attempted to predict human behaviour in theoretical networks based on random probability. Empirical work on stochastic systems has been fuelled by practices in information technology, for example the random access patterns that enable multiple simultaneous mobile telephone use within a community, wireless technologies within a network that initiate and cease randomly, and website hits initiating randomly from across the world (see, e.g., Fenner *et al.* 2006). But despite these models in the quantitative sciences and, ironically, the pervasiveness of postmodern thought in the human sciences, the challenge of examining stochastic human systems in a way that does not focus on continuity (retrospective, invented or otherwise) remains elusive for the human sciences.

One approach for studying stochastic systems emerges from Rabinow's concept of 'event' as a singularity and form. Using the term *form/event*, Rabinow suggested that such a concept could be used to articulate that which makes other things 'more or less suddenly conceivable' (1999, p. 180). A sudden, interrupting telephone call, a rainstorm that forces a group of people into the same restaurant, a meeting on an aeroplane that leads to a new business opportunity are all stochastic forms/events that by chance lead to new aggregations and possibilities. Theoretically, a form/event is stochastic in that it is not indicated as a result of distinct elements involved in its formation but in their randomness. A form/event is not particularly representative of the context or network from which it emerged. Instead, these are new emergences that, in Rabinow's words, 'catalyze previously existing actors, things, temporalities, or spatialities into a new mode of existence, a new asssemblage' (1999, p. 180). Thus, stochastic forms/events are theoretically important for change studies both as interruptions in narrative and as textnologies (strategic change). But the problem remains, how to analytically and empirically account for stochastic events within social networks.

What would an analysis of an event/form look like? Rabinow claimed that forms/events are not quite as mysterious as they would seem. They are able to be analysed and documented; though, he admits that we lack the adequate means to accomplish such work. I have tried to systematize Rabinow's model through the analytical chapters in this book, specifically in my rather stubborn focus on the singularity of the WorkWare campaign as an event and instantiation of change discourse. As I attempted to show, WorkWare, as an event, problematized classifications, practices and things, to use Rabinow's words (1999, p. 181). Within this singularity, events occurred that problematized understandings and classifications of academic governance and the boundaries of corporate influence. These same events confused and re-articulated the then taken-for-granted concepts of change, narrative and transformation, showing them to be enacted strategies with tangential and unpredictable objectives. And, these events also denaturalized taken-for-granted assumptions about leadership, power and influence within what initially was seen as a fairly stable social network on the campus.

I intentionally resisted over-generalizing my findings from the WorkWare study, preferring instead to try and explain why and how the events on the ground during this period occurred. The findings are relative and contingent to this specific social and technological network and I do not claim any reach to or influence on the sort of 'totalizing' categories such as 'culture', 'society', 'science', or even for that matter 'technology', which have come under suspicion in recent approaches to inquiry (Rabinow 1999, p. 181). Thus, technology has been constituted as a specific aggregation of available resources and human objectives within a specific network; change is constituted by its form/event within a specific network (pause in narrative, strategic action, textnology); and politeness is constituted not by universal discourse features, but by the exchange of symbolic currency and self-interest among participants within a specific network. This approach may have application for studies of different projects and sites but what specifically counts for 'change', 'politeness' or 'technology' in this study and in the network examined here is not designed to carry over to other situations without overt comparison and empirical demonstration.

To place a more operational, critical realist frame around the form/event concept, we could turn to various discourse features associated with ambiguity and polysemy as potential markers and components of stochastic systems. As outlined in Chapter 4, we could examine intrinsic and extrinsic modality as polysemous constructions that create potentially (deliberately) random interpretations of social events. These interpretations could then extend in unpredictable ways to enable and instantiate a social change. Other potential features could include clausal and phrasal constructions and the constitution of dynamic (versus synoptic) texts as indicative of relatively unstable and unpredictable (stochastic) discursive/material conditions. Alternatively, we could suggest that forms of elision and textual silence operate as devices that obscure and resist cohesion. These discourse features point to random probabilities, associations and assemblages from which new concepts may emerge.

Without diminishing the descriptive potential of micro-level feature analysis, the discussion of politeness developed in Chapter 6 demonstrated that any examination of discourse, networks and social activity will need to also examine issues of agency and action in order to provide an accurate and robust description of social practice. Recall that a feature analysis alone could not fully answer why people will consent to objectionable or what we called impolite texts. The chapter argued that in some conditions, specific discourse features were secondary characteristics of politeness discourses when other features (self-interest) became more overt and motivational. In these situations, people would trade linguistic currency for self-interest in order to achieve specific social goals. Therefore, we must also consider events and activity when articulating stochastic systems. A working model for such analysis was developed by Spinuzzi, Hart-Davidson and Zachry in their project to model writing processes within organizational networks (2004). Although their model again emphasized the *coordination* of distinct and random elements, they suggested using 'communication event' as a 'primary unit for modeling' (Spinuzzi *et al.* 2004, p. 2). As such, they proposed to consider events (planned and unplanned events: interruptions, conversations, IM, email, meetings, chance encounters) as the rule for written hermeneutics. In this way, such forms/events are not exceptions or distractions from writing but essential components of what is (at least partly) a stochastic process of knowledge invention. Similarly, Hart-Davidson provided a series of interconnected diagrams, including phone calls, face-to-face meetings, interruptions by his daughter, emails, physically typing and time spent reading to illustrate and document communicative events influencing his own writing process within a stochastic network (2004). Removing the necessity of active coordination from these models, we can trace hermeneutics within specific networks as stochastic forms/events leading to new assemblages, products and devices. The act of coordination is simply a retrospective event applied after the fact.

The necessity of power and the limitations of ethics

A study of change must eventually address issues of power within and across the network. At its conclusion, the WorkWare analysis documented the instantiation of a new regime at Northeastern. This new regime was not authoritarian. It was created by technological discursive relations (textnologies). Throughout the campaign, Martin's office provided users with opportunities to dissent. But there was no effective reason or form to effect productive dissent other than simply not adopting the new software. This is different from authoritarianism but it is not purely democracy either. As a textnology, WorkWare was a discourse of power. Thus when OIT 'throws the switch', when they decree that users must use the new servers, their power is egregious. But when they insert themselves, by self-constituted necessity into campus discussions of social and academic relations, they do so through generative power – power they have discursively generated, managed and employ to service their social networks. As I noted in the introduction to this project, I left the WorkWare study somewhat ambivalent about Martin's acts of power. On one level, I understood their exigency, how they benefited Martin and his office, and frankly how a lack of other forms of campus leadership created opportunities for Martin's interventions. At the same time, I continue to be concerned with the use of textnologies to achieve social objectives across the campus, the lack of opportunities for legitimate campus participation in key academic and social issues, and the manipulation of discourse to construct change.

Change studies need to pay careful attention to not only the imposition of power within and across networks, but also to the different ways power is diffused and invested within forms/events of change. Issues of power can inform what I have called studies of the ethics of change, examinations of the ways change is used and instantiated within social networks. But, as my ambivalence above suggests, I am not sure we yet have adequate tools for such a project. Studies examining the ethics of change should be descriptive but also critical as they examine the participation of audiences as networks change, the availability of input and the co-constitution of the new, and the awareness network members have about potential planned results of change. By re-introducing and extending the concept and deployment of discourse technologies, my project has offered new vocabulary and devices for examining the ethics of change. Specifically, the concept of textnologies introduced in this book offers an initial way to define, recognize and limit potentially authoritarian or objectionable methods for instantiating change.

Norman Fairclough concluded his 1992 book *Discourse and Social Change* on alternatively optimistic and pessimistic claims. First, he argued for the inclusion of critical language awareness within language education classes at all levels of education (1992, p. 239). Fairclough's proposal integrated instruction in discourse analysis with detailed examinations of language variation. He also suggested historical and contemporary studies of language standardization, what he called the 'hegemonic processes' that have influenced standard forms (1992, p. 240). To this proposal, we can add studies of social change and examinations of the discursive forms that have constituted and promoted change. In the time since Fairclough's proposal, we have seen considerable development in the field of critical language studies. However, Fairclough's suggestion for more extensive school-based materials for discourse studies remains largely unfulfilled (at least in North America). Now (to employ a textnology) is a particularly significant and appropriate time to advocate (again) for greater analytical attention to discourse within language instruction. Recent testing initiatives in the United States have shown the necessity for renewed attention to language at all levels of education (see, e.g., Fessenden 2006). At the same time, English education initiatives have begun to rediscover the importance and utility of a more structural understanding of language (de Vise 2006). Whether or not we agree with local and national testing initiatives (and if we are comfortable using such a change form/event), we can strategically use results from these projects to advocate for more discourse-based instruction within language classes. In addition, we need to turn the significant methodological discussions of discourse studies toward more initiatives to create curriculum, lesson plans and instructional materials for various levels of education, giving greater attention to specific discourse functions within language instruction.[5]

Fairclough's second, pessimistic, conclusion is more difficult to address. He predicted that the concept of discourse technologies will be problematic for scholars because as more people become aware of the utility of discourse technologies (textnologies), discourse analysis researchers 'will be hard-pressed to prevent their well-intentioned interventions being appropriated by those with the power, resources, and the money' (1992, p. 240). To a degree, Fairclough was correct. As the concept of textnologies has become more widespread we have seen textnologies implemented across professional discourse, institutional settings and political and diplomatic discourses/activities. In its deployment as a change strategy we can certainly document the ways textnologies have colonized other forms/events and how they have been deployed in ever expanding situations. This greater deployment of textnologies again makes the case for a critical realist description of forms/events that instantiate change and an articulation and robust discussion ethics of discourses that facilitate change.

At the same time, despite the expansion of textnologies and their appropriation within a variety of organizational, governmental and professional contexts, there is some sense that they may not be quite as successful as we might predict (or fear). Nearly a decade into the twenty-first century, the activity of planned change remains a problem simultaneously at local and global levels. With few exceptions, relatively high-profile events to alternatively create, impose and attain deliberate changes in organizations, across cultures and within institutions have led to traumatic and rather unsuccessful results. When they have been successful, the influence of textnologies appears temporally limited to short-term and immediate results. Over the longer term, such experiences with planned change have appeared to contradict earlier exuberant claims about the spread of global capitalism, Western values, and market forces predicted in the wake of what was termed the new global economy (Reich 1991; Micklethwait and Wooldridge 2000). Instead, the forms/events designed to instantiate planned social change have appeared to create a renewed emphasis and interest in the local, personal, experiential and in the narrative as an interpretive form. Most dramatically, in its more extreme cases, attempts to instantiate change within and across cultural networks have directly created and fostered forms/events that actively (violently) oppose the new instantiation.

In some ways, these events have constituted a reversal of Fairclough's paradox. Despite two decades of interest, growth and development of discourse studies, there have actually been few models of successful change discourses used seamlessly and without resistance at the sort of scale Fairclough feared. Perhaps this is a good thing. Or, perhaps most efforts, such as the corporate university, simply remain ongoing. To a large extent, discourse studies has been able to remain descriptive and analytical rather than prescriptive, managerial and applied. At the same time, this has not meant that modern leadership practices have abandoned the notion of planned change or the practice of attempting to technologize change. More problematically, despite this interest in change, contemporary events accompanying intentional change practices within and across various networks have shown that Western leadership discourse has yet to consider an ethics of change or the creation of appropriate and accountable methods for instantiating change. Whether or not we agree with these actions, what has become evident in the wake of numerous efforts to intentionally create change is the pressing need for a method and a vocabulary for examining processes of social change and for holding such processes accountable to some workable criteria. Such forms/events would emerge from close, empirical descriptions of change processes; robust but intentionally limited examinations of social change relative to various types of social networks (stochastic, deterministic); case studies of historical and contemporary change initiatives; and ethical investigations into the methods by which change is instantiated and managed. Such a project may not entirely reconstitute the ways change is conducted in modern networks of social, professional, organizational and technical power. But, with some scrutiny and hard questions, it might make us more conscientious about how we conduct change and about the consequences our 'change' may have on others, our organizations and perhaps even ourselves.

Notes

Chapter 1

1 The expansion of corporate and government influence in the university has been well documented. The Cold War and space race funded significant expansion of government and corporate investment in universities, primarily within engineering and science disciplines. Principally, this resulted in patents, technology transfer and other forms of commercialized technology. The growth of global capitalism accelerated these trends and extended corporate interest and the potential for profits into business, the human sciences, and university management and service provision through sub- and direct contracting. See Eurich 1985; Davis and Botkin 1994; Aronowitz 2000; Dolgon 2000; Thompson 2000; White 2000; and Jarvis 2001 for historical and contemporary case studies.

2 As part of my Institutional Review Board agreement, I have used pseudonyms to describe the school, its employees and the people who are mentioned in this book. I have endeavoured to provide sufficient information to enable readers to understand and appreciate the on-the-ground details about the site, the school and the people involved in the study.

3 By cultural-rhetorical, I refer to persuasive acts that are embedded within cultural moments and events. In this specific case, I refer to image and identity as culturally constituted actions. That is, image and identity are formed within and against specific cultural norms and recognitions and the parameters established by the available social network. As cultural constructs, these (and other) embodied actions carry persuasive potential both for the person or organization creating the image or identity and for those who read or engage with that person or organization. For elaboration, see Faber 2002a; Dutton and Dukerich 1981; Dutton *et al.* 1994.

4 The distinction between textual and material becomes complicated toward the latter half of the book. Initially, the distinction is intended to represent activities based largely in discursive production (writing, speaking, visual, semiotics) as textual and non-discursive production (nature, economics, force) as material. However, later parts of the book begin to blur this distinction, especially once text begins to emulate technology, a process I call textnology.

5 There is considerable irony here in that corporate-university consultants and those ostensibly helping universities partner with corporations promoted much of the language and sentiment that marginalized university-based academic sites and practices. See Allen (ed.) 2002.

Chapter 2

1 Proponents of an expanding role for corporate models within academia have not gone unchallenged. However, this challenge has largely come from academics themselves. Aronowitz (2000) argued against corporate education proposals, claiming that the corporate university confuses education with training, grant revenue with knowledge production, and grades and job offers with learning (p. 159). Similarly, Jarvis (2001) wrote that corporate universities are 'education and training establishments that will lead to a form of capitalist cultural reproduction rather than centers of democratic criticality that we might expect from the more traditional university' (p. 128). Corporate models of education have been seen to be responsible for the decline of humanities programs on many campuses (Parenti 2000) and writers have offered ways to resist corporate power on campus, ranging from student-led movements to divest universities from companies that engage in unethical business practices (Ni and Apple 2000) to critiques of campus procurement (Kniffin 2000) and licensing agreements (Benjamin 2000).

2 Perhaps a good initial metaphor to use to introduce discourse strategically designed to accomplish goals is the strategic political tool of gerrymandering. The term dates to 1812 and Elbridge Gerry, a Massachusetts governor who deliberately created election districts that favoured his own political interests. Today, the term is used to identify any strategic redistricting intended to advance political goals. Discursive gerrymandering would entail the deliberate manipulation of discourse to force a constitution of the world onto an (unsuspecting) audience. This manipulation of discourse (and politics) is a technical endeavour so although the metaphor remains appropriate, I use the already established term 'technologized discourse' and in the book's conclusion 'textnologies' to refer to this practice.

3 Krause (1996) argued that the professions, the state and capitalism are situated in structural tension within Western economies. This tension acts as a sort of structural check on the powers of each sector. As capitalist practices become egregious, the professions and the state begin to intervene. Similarly, if the professions or the state are perceived to be too powerful, the other sectors will check their power. Krause's description of the professions is limited to core professions: medicine, law, clergy, academia, and to some degree applied and consulting sciences (in Europe).

4 Close readers will recognize various moments in the book where my own discourse emulates the technocratic style – 'decisions are made'. These examples are deliberately employed to demonstrate the technique.

5 Peter Drucker's work has been influential in positioning training and human resources as essential components of organizational strategy; see, e.g., Drucker 1993 for a description of training as a management technique.

Chapter 3

1 Krug's position is critical of the apparent non-discursive nature of technology. He argued that technology disperses language and the meaning function of language. He wrote 'technology dramatically alters our understanding of the world both by becoming the object of the world and through assimilating language into its goals and values' (2005, p. 19). Further, 'With technology, once

the inadequacy of traditional, normative language is established, one has no alternative but to chase after technology, seeking to experience through it. We find the technological gaze on television, the periscope, the microscope, the CAT scan; and increasingly the image defines what we are' (p. 21). While my description admittedly characterizes two extreme positions, Krug's own position of technology out of control is reflective of considerable work in the human sciences. See, e.g., Habermas 1987. For this tradition of scholarship, mining is the model for the genesis of technology. As Krug put it, technology was born from exploitation, 'their applications exemplified the same disregard for both human beings and the environment' (p. 14).

2 Latour's assertion that 'technology gives itself a context' is disputed. Johnson-Eilola argues technology can not simply provide its own network/context because it is usually a contested articulation, 'one in which forces and subjects within the context make material and rhetorical moves to push specific, different articulations' (email correspondence).

3 As one member of the advisory committee noted to me, it was extraordinary that OIT was endeavouring to implement WorkWare plus a comprehensive data management system for financials, student records and course scheduling and an online instructional system at the same time.

4 Some members of the advisory committee claim that this may not be accurate. However, this is a disputed point. According to the diverging perspective, OIT agreed (decided) to use the beta version because of the cost savings associated with it. In addition, it would make more sense to install the newest version of WorkWare rather than install an older version and then pay for the upgrade. The advisory committee did not know that the selected version of WorkWare was beta until Martin approached the committee in the middle of the implementation and explained that the difficulties were because he had agreed to install the beta version for financial reasons. In addition, there is some question as to the extent to which Martin realized or knew that the beta version would be so unstable.

Chapter 4

1 The theme of discourses of change not emphasizing continuity will be further developed in the concluding chapter which contrasts differing approaches to change.

2 'Identical discourses that reoccur at temporal intervals, but not as linked, progressive, or incremental stages, suggests a different relationship between discourse and social transformation' would seem to be a useful definition for a study of intertextuality and its function (or ability) to spark and manage social change. Chapter 5 specifically addresses intertextuality. Yet, I would like to emphasize that my analysis in these chapters is not designed to provide a universal grammar for understanding or implementing change but a description of the way change occurred within these texts. In some cases, my analysis reinforces variables identified in other studies. In other cases, my analysis identifies relatively new features of change discourse. The point here is to identify features in the hope that other studies will test and examine similar texts in a continued endeavour to understand the various ways change can function.

3 There is some added complexity here (which I return to in Chapter 7) in that OIT could easily tell who was using the new system by simply tracing user

identification logs on the new servers against the accounts held by people on campus. As a consequence, they could have simply contacted these people directly – probably in person or by telephone since it would be obvious that they are not using their email. This is not so much a problem for the clausal analysis here (since the clause remains structurally dynamic), but does problematize the interpretation why the campaign is dynamic at this late date. As I note in Chapter 7, one rationale is that OIT is managing the perceptions of those who have already switched to the new servers. As such, the dynamic style would be indicative of OIT's inability to fully understand or measure audience support for the initiative.

4 See note 3 above.

5 Unfortunately, this has already been a response to early presentations of this data. One audience member approached me after a presentation to say that he was going back to his office to 'entirely revise' an email message he was composing before he left to attend my talk.

Chapter 5

1 To some extent, it should be noted that the observer quite possibly tainted the observation in these meetings. I did observe Martin integrating these themes in meetings I attended – both with his staff and exclusively with the two of us. But these were audiences who were comfortable with this discourse. In other meetings, for example with the advisory committee and in campus-wide forums, Martin's thematic scope was not as broad as it was in these more private networks and situations. On management lessons from Shackleton, see Morrell and Capparell 2001. Martin used this text in a management study group in his office.

2 The issue of volition with the user name convention is more evident and contested than I have previously narrated. One member of the advisory committee told me that at several meetings, OIT personnel insisted that students and faculty/staff needed different formats for their user IDs. OIT personnel said this was important because they wanted to be able to differentiate student accounts from faculty and staff accounts. There was no reason given for why this distinction was important or necessary. In the end, faculty and staff were able to retain previous identifications (nicknames, just last name without first initials). New accounts for faculty and staff though were created with first initial(s) and last name (bfranklin) while new accounts for students were last name then first initial (franklinb). This small example demonstrates that the ultimate decision was not technical but created and implemented because of human preference.

3 The angels and demons theme was inspired by the summer 2006 Standing Conference on Organizational Symbolism, The Good the Bad and the Ugly: Organizations and Demons hosted by Radboud University in Nijmegen, The Netherlands. An early version of sections of this chapter was presented there.

4 My claim here is an intertextual referent to a rather well known New Testament teaching that Christians should be *in* the world but not *of* the world (Romans 12:2; John 17:6–19). Ironically, this is the ultimate subject-position for a management consultant.

Chapter 6

1 This claim has often resulted in discussions about intentionality – do writers and speakers consciously create (manipulate) discourse to form technologized texts. The answer is relevant but probably situation-dependent. To an extent, Martin was conscious of his own technologized discourse, though he didn't refer to it this way. At the same time, as these practices become more practised and ingrained within everyday events, people will not realize when they employ a technologized discourse – much the same way as few people will admit to deploying pathos to jump the queue when running late to buy a coffee even if they recognize the rhetorical technique being used by others.

2 Considerable current public discourse occurs at the level of content with few people examining the simple structures of discourse or argument. The most significant problem with this feature is that quite often the 'table is set' before the debate even begins. 'A forum about effective WorkWare implementation' necessarily excludes talk about other platforms, discussion about the necessity of WorkWare itself, or suggestions for improving IT client services. Of course, contesting the forms in which discourse is presented to us is a much more difficult and problematic endeavour and we do not have ample resources for such a task, especially in the problematic context of public discourse. However, that is not to say that this does not occur. A considerable problem becomes anticipating such activities to ensure that one can adopt more than simply resistant or reactive stances.

3 Paul Rabinow has criticized what he calls the 'over proliferation of totalizing systems' within the human sciences (2003, pp. 29–30). This criticism is taken up in my conclusion (Chapter 7) along with Rabinow's proposal to establish theoretically informed empirical studies of human activity. Rabinow's approach is not anti-theory but seeks to integrate theoretical understanding into empirical work. Such an approach could be seen as an empirical study of concepts, or the study of enacted theory.

4 On the concept of elite knowledge, professional power and professional communication, see Grabill 2001; Faber 2002b. On professional power derived from restricting knowledge, see Bledstein 1976; Larson 1977; Friedson 1986.

5 Simpson noted that the rationalization of the life world is 'an ironic and problematic process' for Habermas (1995, p. 85). It is ironic because it is exactly the 'spirit of modernity' and its analytical and critical imperative, which Habermas celebrates, that advocates the disenchantment and fracturing of tradition. It is problematic because it puts again into dispute those scaffolds that supported a pre-technological perspective, effectively not allowing those dispositions to have adequate currency or stability to offset the values of technocratic orientations. See discussion in Simpson 1985, pp. 84–6.

6 Scollon noted that there is some ambiguity in the concept of 'mediational means' as some are physical objects while others are more elusive and abstract. See discussion in Scollon 2001, p. 120.

7 Watts' argument, to restore original claims and approaches to the study of politeness, an argument from precedent, is a relatively common change narrative itself. The argument form asserts that because something was done effectively in the past, but has for some reason been marginalized or no longer enacted in the present, it should be reenacted in the future. For example in Faber 2002a, I documented a change management case at a bank that was based

on the mantra 'we have always been adept at responding to our customers' needs but our current practices no longer put the customer's needs first'. Therefore, the bank employees were required to change their day-to-day activities (sell more profitable stocks, bonds, and mutual funds). The fact that these services were unavailable when the bank was 'responding to customers' needs' in the past was not important.

8 There is another dynamic occurring here that Stan Deetz has discussed in his work on public forums (Deetz and Irvin 2006). Deetz argued that in an American context, the function of a public forum (public hearing) is to allow members of a community to voice their opinions about a specific project or policy. Members participate to ensure that their perspective is heard and those conducting the meeting place a priority on making sure that participants know that their position was indeed voiced. The public meeting, however, is not designed to create policy, achieve collaboration or move policies towards some sort of middle ground (compromise). This is understood and anticipated by most participants.

9 Data cited from Northeastern Tech's student newspaper coverage about the campus meeting held to discuss VOIP in the student dormitories.

Chapter 7

1 I am borrowing the term form/event from Rabinow (1999, p. 180). In other places, I have used the term meaning/event to describe a similar enactive process of generative construction (Faber 2007). 'Form' provides a useful link to the structural analysis of the previous chapters and 'meaning' can be seen to be integrated within 'form'.

2 Harvey noted two considerable problems with this notion of the residual as an enabling force. First, it seems primarily driven to solve theoretical problems rather than actual on-the-ground occurrences. Second, it appears to constitute the residual as the essential and sole cause of change itself when we know that change can be caused by a variety of events, practices and material occurrences.

3 I should stress here that I am not arguing that the Classical conception of change is no longer held or privileged within many social networks. Classical accounts of change continue to inform human experiences and networks and numerous authors critical of (post)modern social relations have attempted to restore the value of continuity, stability and narrative within contemporary social systems. Thus, I am not describing or theorizing a chronological process in which one perception of change transitions into another. The problem I have tried to study in this book has been the emergence of change as a technology within specific, key networks and the implication of this emergence for organizational life, communication and persuasion, and the question of ethics. Thus, what I am articulating is perhaps best seen as a dynamic and a struggle among interpretations of and strategies for social exigence.

4 Citing Hans Blumenberg, Rabinow's description was less kind. He argued that 'to that extent the philosophy of history is an attempt to answer a medieval question with the means available to a post medieval age' (Rabinow 2003, p. 29).

5 The field of discourse studies has produced numerous overlapping introductions, methodology texts and other such guidebooks largely addressed to the university audience. These certainly fulfil a market need and have helped to promote the field within undergraduate and graduate programmes. What the

field has not done is to develop similar materials for earlier educational experiences. A model for such curriculum development can be found in the numerous STEM (Science, Technology, Engineering, and Math) initiates promoted by the National Science Foundation in the United States. These initiatives connect STEM programmes in universities with class work, assignments and projects in kindergarten to grade 12 courses across the country. They also promote summer institutes and programmes that bring students to university campuses and create active, project-based learning opportunities for students.

Glossary

Action processes. Accent accomplishments and realize processes: e.g. 'we manage change'. Action processes are significant because they constitute the 'noun' as real and situate the verb phrase as a concrete process rather than a possibility. For example, the phrase 'manage change' implies that change is real, it has already happened, and places it within an ongoing management process. The action process could be seen as a device for presupposition since negating the phrase does not negate the existence of the noun: 'do not manage change' → the negation does not influence the existence of change.

Adverbial. Functional unit of discourse, usually a phrase. Adverbials modify a verb by describing location, goal, time, frequency, manner, reason and purpose. Adverbials can occur as prepositional phrases, adverb phrases or noun phrases and can be placed in various locations throughout the sentence (beginning, middle or end) without influencing sentence meaning.

Amplification. Structuring discourse to highlight certain facts or processes.

Copula verbs and subject complements. In this construction, the subject complement (noun phrase following copula verb) redefines the initial subject providing a simple transition from one concept to the other and over the length of the clause from early mention, to redefinition, to instantiation. The copula effectively reconstructs the local context enabling the writer to build a 'new' from an existing 'old'.

Determiners. The and A. '*The*' refers to a referent with which the speaker/writer presumes the audience is already familiar. However, the assumption of familiarity does imply a certain amount of estimation and can be used to impose the fiction of existing information onto an unsuspecting audience. Unlike the definite article 'the', the indefinite 'a' refers to new information, information that is assumed to be first mention.

Deixis. Cohesion patterns, typically with reference to entities outside the text (see also paradigmatic references).

Dynamic style. Highly claused, positions utterances as less stable. Opposite of synoptic style.

Genre. Habitual and recognized forms/events of discourse, can include speech, writing, electronic text, music. Genres are both discursive and social processes.

Given information/new information. 'Given' refers to information the speaker or writer presumes the audience already knows (is familiar with) while 'new' refers to information that is assumed to be new information.

Grammatical metaphor. See nominalization. Grammatical metaphor refers to one grammatical type taking on the construction of another. For example a verb taking on the properties of a noun (nominalization).

Hedge. Words that reduce the strength of commitment to a claim: such, like, could, many, some. Hedges also substitute for concrete references and can create ambiguity about the precision of a claim.

Hybridity. Co-construction of two or more genres into a new form/event. A 'hybrid genre'.

Ideational level. The level of semantic meaning (content). The construction of systems, knowledge and belief (Fairclough 1992, p. 64).

Interpersonal level. Roughly aligned with pragmatic level and the study of cohesion. The level at which the text constructs identifications and relationships between and among people. Also called relational (Fairclough 1992, p. 64).

Intertexuality. 'The property texts have of being full of snatches of other texts, which may be explicitly demarcated or merged in, and which the text may assimilate, contradict, ironically echo, and so forth' (Fairclough 1992, p. 84). Intertextuality makes present an absent alternative narrative through semantic reference, symbolic use or even structural imitation.

Metaphor. Comparison between two objects that directly constitutes the one object as the other.

Modality. The auxiliary verbs can, could, would, should, shall, may, might, will and must. Modality can be a key feature of destabilizing discourses (Faber 2003) and modals are unique from other verbs in that they express a degree of conditionality. For example, the phrase *Mary can swim across the river* expresses different meaning from *Mary will swim across the river,* or *Mary is swimming across the river.* Functionally, modality can be characterized in two ways, intrinsic and extrinsic modality (Quirk *et al.* 1985, p. 219). Intrinsic modality (root modality) describes permission, obligation and volition. These concepts identify and complicate human control over specific events. Extrinsic (epistemic) modality refers to necessity, prediction or possibility.

Nominalization. Resituating a verb as a noun: 'The implementation of WorkWare' rather than 'implementing WorkWare'. Nominalization can be used to avoid assigning agency to a claim or action.

Paradigmatic. Referring to words/concepts outside of the text. Opposite of syntagmatic.

Passivization. Use of a passive verb: 'a new system was created'. Passivizations are important because they remove agency from the phrase to create the appearance of actions that occur without or beyond human intervention.

Presupposition. An assumed meaning embedded within an utterance: 'Are you still angry with your room-mate?' Fairclough wrote about presuppositions in three ways, existential assumptions (assumptions about what exists), propositional assumptions (assumptions about what is, can be or will be) and value assumptions (what is good, desirable or bad) (Fairclough 2003, p. 55). In each case, a presupposition evokes a claim as shared understanding in a way that makes it difficult to challenge the verity of that claim. Presuppositions can be revealed by negating the claim. What is left behind is what has been presupposed.

Progressive aspect. Verb type (is/are __ing) related to the functional, strategic role played by modality (Faber 2003). Unlike traditional forms of modality, the progressive aspect focuses on time (they **are playing** hockey; she **is volunteering** for the hospital; corporate universities **are becoming** a competitor in the higher education marketplace). In these cases, the progressive aspect hedges the exact reference to when an activity started and ended. This makes it impossible to adequately evaluate the temporal duration of claims like '*corporate learning centers* ***are evolving*** *into universities*, or *education* ***is becoming*** *a scarce commodity*'. See also temporal ambiguity.

Pronouns. Words that substitute for nouns and noun phrases: it, they, him, her, your. They can be used to create ambiguity when the substitution is vague or imprecise.

Stabilization/destabilization. Stability is a rough estimation of the confidence writers (or speakers) have in their ability to make a claim. Stability refers to the network environment into which a text is situated, the amount of lexical work a speaker or writer must accomplish to orient the audience to a claim, the amount of background information needed to support meaning and other contextual issues that impede understanding. As claims require further and further modification (or explanation) they become more and more unstable. As such, the dynamic style is indicative of a changing, unstable network.

Strategic ambiguity. An intentionally destabilizing claim that cannot be immediately resolved or solidified by available textual or contextual information.

Synoptic style. Fewer clauses, positions utterances as more stable. Opposite of dynamic style.

Syntagmatic. Referring to words/concepts within the text. Opposite of paradigmatic.

Temporal ambiguity. Using verb types that constitute time to hedge details about action: 'Problems are beginning to emerge'.

Textual level. Also referred to as syntax. The level of individual grammatical units.

Theme/rheme. 'Theme' refers to the initial reference of a clause (generally what the clause is about), typically the semantic content before the verb. 'Rheme' typically refers to everything after the verb. See also given and new.

References

Allen, M. (ed.) (2002) *The Corporate University Handbook: Designing, Managing, and Growing a Successful Program.* New York: AMACOM.

Alvesson, M. and Kärreman, D. (2000) 'Taking the linguistic turn in organizational research'. *Journal of Applied Behavioral Science*, 36, 136–58.

Aristotle (1991) *On Rhetoric: A Theory of Civic Discourse* (ed. and trans. by George Kennedy). Oxford: Oxford University Press.

Aronowitz, S. (2000) *The Knowledge Factory: Dismantling the Corporate University and Creating True Higher Learning.* Boston: Beacon Press.

Bakhtin, M. (1986) *Speech Genres and Other Late Essays* (C. Emerson and M. Holquist, eds, V. McGee, trans.). Austin: University of Texas Press.

Bartlett, T. (2004) 'Mapping distinction: Towards a systemic representation of power in language', in L. Young and C. Harrison (eds), *Systemic Functional Linguistics and Critical Discourse Analysis: Studies in Social Change.* London: Continuum, pp. 68–84.

Baudrillard, J. (1983) *Simulations.* New York: Semiotext(e).

Bazerman, C. (1988) *Shaping Written Knowledge: The Genre and Activity of the Experimental Article in Science.* Madison: University of Wisconsin Press.

— (1999) *The Languages of Edison's Light.* Cambridge, MA: MIT Press.

— (2001) 'Nuclear information. One rhetorical moment in the construction of the information age'. *Written Communication*, 18, 259–95.

Bazerman, C. and Paradis, J. (eds) (1991) *Textual Dynamics of the Professions: Historical and Contemporary Studies of Writing in Professional Communities.* Madison: University of Wisconsin Press.

Benjamin, M. (2000) 'Toil and trouble: Student activism in the fight against sweatshops', in G. D. White (ed.), *Campus Inc.: Corporate Power in the Ivory Tower.* Amherst, NY: Prometheus, pp. 237–52.

Berkenkotter, C. and Huckin, T. (1995) *Genre Knowledge in Disciplinary Communication: Cognition/Culture/Power.* Hillsdale, NJ: Lawrence Erlbaum.

Bledstein, B. (1976) *The Culture of Professionalism: The Middle Class and the Development of Higher Education in America.* New York: Norton.

Bloom, H. (1988) *Closing of the American Mind.* New York: Simon & Schuster.

Bloor, T. and Bloor, M. (2003) (reprint) *The Functional Analysis of English: A Hallidayan Approach.* London: Arnold. Original edition, 1995.

Bourdieu, P. (1977) *Outline of a Theory of Practice* (R. Nice, trans.). Cambridge: Cambridge University Press.

— (1990) *The Logic of Practice* (R. Nice, trans.). Sanford: Sanford University Press.

— (1991) *Language and Symbolic Power* (J. Thompson, ed., G. Raymond and M. Adamson, trans.). Cambridge, MA: Polity.

Bourdieu, P. and Wacquant, L. (1992) *An Invitation to a Reflexive Sociology*. Chicago: University of Chicago Press.

Boyce, D. (1996) 'Organizational story and storytelling: A critical review'. *Journal of Organizational Change Management*, 9, 5–26.

Brown, P. and Levinson, S. (1978) *Politeness: Some Universals in Language Use*. Cambridge: Cambridge University Press.

Brown, G. and Yule, G. (1983) *Discourse Analysis*. Cambridge: Cambridge University Press.

Burke, K. (1963) *Language as Symbolic Action: Essays on Life, Literature, and Method*. Berkeley: University of California Press.

— (1969 [1950]) *A Rhetoric of Motives*. Berkeley: University of California Press.

Burrell, G. (1988) 'Modernism, postmodernism and organizational analysis 2: The contribution of Michel Foucault'. *Organization Studies*, 9, 221–35.

Cartwright, L. (1995) *Screening the Body: Tracing Medicine's Visual Culture*. Minneapolis: University of Minnesota Press.

Case, P. (1999) 'Remember re-engineering? The rhetorical appeal of a managerial salvation device'. *Journal of Management Studies*, 36, 421–40.

Chilton, P. (2004) *Analysing Political Discourse: Theory and Practice*. London: Routledge.

Chouliaraki, L. and Fairclough, N. (1999) *Discourse in Late Modernity: Rethinking Critical Discourse Analysis*. Edinburgh: University of Edinburgh Press.

Codrescu, A. (2005) (4 May) 'The reckless folly of youth'. Broadcast on National Public Radio. Available online at www.npr.org/templates/story/story.php?storyId=4630857 (10 May 2005).

Connell, I. and Galasiñski, D. (1998) 'Academic mission statements: An exercise in negotiation'. *Discourse and Society*, 9, 457–79.

Cook, S. and Suter, R. W. (1980), *The Scope of Grammar: A Study of Modern English*. New York: McGraw-Hill.

Davis, S. and Botkin, J. (1994) The *Monster Under the Bed*. New York: Touchstone.

Deetz, S. and Irvin, L. (2006) 'Discourses of democracy and forms of contestation in collaborative processes'. Paper presented at Organizational Discourse: Identity, Ideology, and Idiosyncrasy. Vrije Universiteit, Amsterdam, 28 July 2006.

Deleuze, G. and Guattari, F. (1987) *A Thousand Plateaus, Capitalism and Schizophrenia* (B. Massumi, trans.). Minneapolis: University of Minnesota Press.

Denzin, N. K. (1991) *Images of Postmodern Society*. Newbury Park, CA: Sage.

Derrida, J. (1988) 'Signature Event Context', in G. Graff (ed.) *Limited Inc*. Evanston IL: Northwestern University Press.

de Vise, D. (2006) (23 October) 'Clauses and Commas make a Comeback: SAT helps Return Grammar to Class'. *Washington Post*, p. A01.

Disterheft, D. (2004) *Advanced Grammar: A Manual for Students*. Upper Saddle River, NJ: Pearson.

Dolgon, C. (2000) 'Justice for janitors: Organizing against outsourcing at Southhampton College'. G. White with F. Hauck (eds), *Campus Inc.: Corporate Power in the Ivory Tower*. Amherst, NY: Prometheus Books, pp. 342–57.

Drucker, P. (1993) *Post Capitalist Society*. New York: Harper.

Dutton, I. E. and Dukerich, J. M. (1981) 'Keeping an eye on the mirror: Image and identity in organizational adaptation'. *Academy of Management Journal*, 34, 517–54.

Dutton, I. E., Dukerich, J. M. and Harquail, C. V. (1994) 'Organizational images and member identification'. *Administrative Science Quarterly*, 39, 239–63.

Eco, U. (September 1994) 'La bustina di Minerva'. *Espresso*, 30: back page.

Ellul, J. (1964) *The Technological Society* (J. Wilkinson, trans.). New York: Alfred A. Knopf.

Engeström, Y. (1990) *Learning, Working, and Imagining: Twelve Studies in Activity Theory*. Helsinki: Orienta-Konsultit Oy.

Eurich, N. (1985) *Corporate Classrooms: The Learning Business*. Princeton: The Carnegie Foundation for the Advancement of Teaching.

Evans, S. (2005) (21 April) 'Apple a day keeps the music at play'. BBC News. Available online at www.news.bbc.co.uk/1/hi/programmes/from_our_own_correspondent/4464735.stm (17 August 2006).

Faber, B. (2002a) *Community Action and Organizational Change: Image, Narrative, Identity*. Carbondale: Southern Illinois University Press.

— (2002b) 'Professional identities: What is professional about professional communication?' *Journal of Business and Technical Communication*, 16, 306–37.

— (2003) 'Creating rhetorical stability in corporate university discourse: Discourse technologies and change'. *Written Communication*, 20, 391–425.

— (2006) 'Representations of nanoscience and technology in popular media'. *Technical Communication Quarterly*, 15, 141–69.

— (2007) 'Discourse and regulation: Critical text analysis and workplace studies', in M. Zachry and C. Thralls (eds), *Communicative Practices in Workplaces and the Professions: Perspectives on the Regulation of Discourse and Organizations*. Amityville, NY: Baywood, pp. 203–18.

Fairclough, N. (1989) *Language and Power*. London: Longman.

— (1992) *Discourse and Social Change*. Cambridge: Polity.

— (1995) *Critical Discourse Analysis: The critical study of language*. London: Longman.

— (1996) 'Technologization of discourse', in R. Caldas-Coulthard and M. Coulthard (eds), *Texts and Practices: Readings in Critical Discourse Analysis*. London: Rout-ledge, pp. 71–83.

— (2003) *Analyzing Discourse: Textual Analysis for Social Research*. London: Routledge.

— (2004) 'Critical discourse analysis in researching language in the new capitalism: Overdetermination, transdisciplinarity, and textual analysis', in L. Young and C. Harrison (eds), *Systemic Functional Linguistics and Critical Discourse Analysis: Studies in Social Change*. London: Continuum, pp. 103–22.

— (2005), 'Discourse analysis in organization studies: The case for critical realism'. *Organization Studies*, 26, 915–39.

Fenner, T., Levene, M., Loizou, G. and Roussos, G. (2006) 'A stochastic evolutionary growth model for social networks'. *Physics Abstract*. Available online at www.arxiv.org/abs/physics/0607188 (October 12 2006).

Fessenden, F. (2006) (1 October) 'In English, gaps between grades'. *New York Times*, Section 14WC, p. 8, column 2.

Feyerabend, P. (1991) *Three Dialogues on Knowledge*. Oxford: Basil Blackwell.

Foucault, M. (1970) *The Order of Things: An Archaeology of the Human Sciences*. New York: Vintage Books.

— (1972) *The Archaeology of Knowledge and The Discourse on Language*. (A. M. Sheridan Smith, trans.). New York: Pantheon.

— (1973) *Madness and Civilization: A History of Insanity in the Age of Reason* (R. Howard, trans.). New York: Vintage.

— (1978) *The History of Sexuality: An Introduction* (R. Hurley, trans.). Harmondsworth: Penguin.

— (1979a) *Discipline and Punish: The Birth of the Prison* (A. Sheridan, trans.). New York: Vintage.
— (1979b) 'Omnes et Singulatim: Towards a criticism of political reason. The Tanner Lectures on Human Values,' delivered at Stanford University, 10 and 16 October 1979. Available online at www.foucault.info/documents/foucault.omnesEtSingulatim.en.html
— (1986) 'Of other spaces'. *Diacritics* 16 (Spring 1986) 22–7.
— (1988) 'Politics and reason,' in L. Kritzman (ed.), *Michel Foucault Politics Philosophy Culture: Interviews and Other Writings 1977–1984.* New York and London: Routledge.
— (1994) *The Birth of the Clinic: An Archaeology of Medical Perception* (A. Sheridan Smith, trans.). New York: Vintage.
— (2003) *Society Must Be Defended: Lectures at the College de France, 1975–76* (D. Macy, trans.). New York: Picador.
— (2003b) *Abnormal: Lectures at the College de France, 1974–75* (G. Burchell, trans.). New York: Picador.
Foulkes, P. and Docherty, G. (2005) 'The social life of phonetics and phonology'. *Journal of Phonetics.* Article available online at www.users.york.ac.uk/~pdf11/jphon06.pdf
Fowler, R. (1996) 'On critical linguistics,' in C. R. Caldas-Coulthard and M. Coulthard (eds), *Texts and Practices: Readings in Critical Discourse Analysis.* London: Routledge, pp. 3–14.
Fowler, R., Hodge, R., Kress, G. and Trew, T. (eds) (1979) *Language and Control.* London: Routledge and Kegan Paul.
Friedson, E. (1986) *Professional Powers: A Study of the Institutionalization of Formal Knowledge.* Chicago: University of Chicago Press.
Gabriel, Y. (1998) 'Same old story or changing stories? Folkloric, modern and postmodern mutations,' in D. Grant, T. Keenoy and C. Oswick (eds), *Discourse and Organization.* London: Sage, pp. 84–103.
Galbraith, J. (1968) *The New Industrial State.* New York: The New American Library.
Gee, J. P. (1999) *An Introduction to Discourse Analysis: Theory and Method.* London: Routledge.
Gee, J., Hull, G. and Lankshear, C. (1996) *The New Work Order: Behind the Language of the New Capitalism.* Boulder, CO: Westview.
Gibson, W. (1984) *Neuromancer.* New York: Ace Books.
Giddens, A. (1984), *The Constitution of Society: Outline of the Theory of Structuration.* Berkeley: University of California Press.
— (1991) *Modernity and Self-Identity.* Cambridge: Polity Press.
Giltrow, J. (1998) 'Modernizing authority: Management studies and the grammaticalization of controlling interests'. *Journal of Technical Writing and Communication,* 28, 265–86.
Giroux, H. (1988) *Schooling and the Struggle for Public Life: Critical Pedagogy in the Modern Age.* Minneapolis: University of Minnesota Press.
— (1992) *Border Crossings: Cultural Workers and the Politics of Education.* New York: Routledge.
Goffman, E. (1982) *Interaction Ritual.* New York: Pantheon.
Goodall, H. L. (1994) *Casing a Promised Land: The Autobiography of an Organizational Detective as Cultural Ethnographer.* Carbondale and Edwardsville: Southern Illinois University Press.
Grabill, J. (2001) *Community Literacy Programs and the Politics of Change.* Albany: SUNY Press.

Graham, P. (2004) 'Predication, propagation, and mediation: SFL, CDA, and the inculcation of evaluative-meaning systems', in L. Young and C. Harrison (eds), *Systemic Functional Linguistics and Critical Discourse Analysis: Studies in Social Change.* London: Continuum, pp. 52–67.

Grant, D., Keenoy, T. and Oswick, C. (1998) 'Introduction: Organizational discourse: Of diversity, dichotomy, and multi-disciplinarity', in D. Grant, T. Keenoy and C. Oswick (eds), *Discourse and Organization.* London: Sage, pp. 1–13.

Grice, H. P. (1975) 'Logic and conversation', in P. Cole and J. L. Morgan (eds), *Syntax and Pragmatics 3: Speech Acts.* New York: Academic Press, pp. 41–58.

— (1978) 'Further notes on logic and conversation', in P. Cole (ed.), *Syntax and Semantics 9: Pragmatics.* New York: Academic Press, pp. 113–28.

Habermas, J. (1987) *The Theory of Communicative Action, Vol 2* (T. McCarthy, trans.). Boston: Beacon Press.

Hall, S. (1989) 'Ideology and communication theory', in L. Grossberg, B. Dervin, B. O'Keefe and E. Wartella (eds), *Rethinking Communication: Vol 1 Paradigm Issues.* Newbury Park: Sage, pp. 40–52.

Halliday, M. A. K. (1967) 'Notes on transitivity and theme in English: Part 2'. *Journal of Linguistics* 3, 199–244.

— (1994) *An Introduction to Functional Grammar* (2nd edn). Edward Arnold: London.

Halliday, M. A. K. and Hasan, R. (1976) *Cohesion in English.* London: Longman.

Halliday, M. A. K. and Martin, J. (1993) *Writing Science: Literacy and Discursive Power.* Pittsburgh: University of Pittsburgh Press.

Halliday, M. A. K. and Matthiessen, M. I. M. (2004) *An Introduction to Functional Grammar* (3rd edn). London: Arnold.

Hammer, M. and Champy, J. (1993) *Reengineering the Corporation: A Manifesto for Business Revolution.* New York: HarperBusiness.

Hanna, D. E. (ed.) (2000) *Higher Education in an Era of Digital Competition: Choices and Challenges.* Madison, WI: Atwood Publishing.

Hardy, C., Lawrence, T. and Phillips, N. (1998) 'Talk and action: Conversations and narrative in interorganizational collaboration' in D. Grant, T. Keenoy and C. Oswick, *Discourse and Organization.* London: Sage, pp. 65–83.

Harrison, C. and Young, L. (2004) 'Bureaucratic discourse: Writing in the "comfort zone" ', in L. Young and C. Harrison (eds), *Systemic Functional Linguistics and Critical Discourse Analysis: Studies in Social Change.* London: Continuum, pp. 231–46.

— (2005) 'Leadership discourse in action: A textual study of organizational change in a Government of Canada department'. *Journal of Business and Technical Communication,* 19, 42–77.

Hart-Davidson, W. (2004) 'Reconsidering the basics of Technical Writing: What are the basic units of a quality writing project?' Paper presented at the Council for Programs in Technical and Scientific Communication. Clarkson University, October 2004. Paper available online at www.msu.edu/hartdav2/cptsc2004.pdf (25 October 2006).

Harvey, A. (2004) 'Charismatic business leader rhetoric: From transaction to transformation', in L. Young and C. Harrison (eds), *Systemic Functional Linguistics and Critical Discourse Analysis: Studies in Social Change.* London: Continuum, pp. 231–46.

Harvey, D. (1996) *Justice, Nature and the Geography of Difference.* London: Blackwell.

Hasan, R. (1999) 'The disempowerment game: A critique of Bourdieu's view of language'. *Linguistics and Education*, 10, 443–59.

— (2004) 'Analysing discursive variation', in L. Young and C. Harrison (eds), *Systemic Functional Linguistics and Critical Discourse Analysis: Studies in Social Change*. London: Continuum, pp. 15–52.

Hawthorne, E. M., Libby, P. A. and Nash, N. S. (1983) 'The emergence of corporate colleges'. *Journal of Continuing Higher Education*, 2, Higher Education Publications Inc., pp. 2–9.

Heath, R. (1994) *Management of Corporate Communication: From Interpersonal Contacts to External Affairs*. Hillsdale, NJ: Lawrence Erlbaum.

Hein, K. (September 1999) 'Class culture.' *Incentive* 173, 75–8.

Held, G. (1992) 'Politeness in linguistic research' in R. Watts, S. Ide and K. Ehlich (eds), *Politeness in Language: Studies in its History, Theory, and Practice*. Berlin: Mouton de Gruyter, pp. 131–53.

Hodge, R. and Kress, G. (1993) *Language as Ideology* (2nd edition). New York: Routledge.

Holden, A. and Elden, S. (2005) 'It cannot be a real person, a concrete individual: Althusser and Foucault on Machiavelli's political technique'. *Borderlands E-Journal* 4, available online at www.borderlandsejournal.adelaide.edu.au/vol4no2_2005/eldenhold_foucault.htm. (27 October 2006).

Huckin, T. (1992) 'Context-sensitive text analysis' in G. Kirsch and P. Sullivan (eds), *Methods and Methodology in Composition Research*. Carbondale and Edwardsville: Southern Illinois University Press, pp. 84–104.

— (2002) 'Textual silence and the discourse of homelessness'. *Discourse and Society*, 13, 347–72.

— (2004) 'Content analysis: What texts talk about', in C. Bazerman and P. Prior (eds), *What Writing Does and How it Does it: An Introduction to Analyzing Texts and Textual Practices*. Mahwah, NJ: Lawrence Erlbaum Associates, pp. 13–32.

Hull, D. (1988) *Science as a Process: An Evolutionary Account of the Social and Conceptual Development of Science*. Chicago: University of Chicago Press.

Jameson, F. (1991) *Postmodernism, or the Cultural Logic of Late Capitalism*. Durham, NC: Duke University Press.

Jarratt, S. (1992) *Rereading the Sophists: Classical Rhetoric Refigured*. Carbondale and Edwardsville: Southern Illinois University Press.

Jarvis, P. (2001) *Universities and Corporate Universities: The Higher Learning Industry in a Global Society*. London: Kogan Page.

Johnson, R. (1998) *User-Centered Technology: A Rhetorical Theory for Computers and other Mundane Artifacts*. Albany: SUNY Press.

Johnson-Eilola, J. (1997) *Nostalgic Angels: Rearticulating Hypertext Writing*. Stamford, CT: Ablex.

— (2005) *Datacloud: Toward a New Theory of Online Work*. Cresskill, NJ: Hampton Press.

Katz, S. (1992) 'The ethic of expediency: Classical Rhetoric, technology, and the Holocaust'. *College English*, 54, 255–75.

Kelly-Holmes, H. (1998) 'The discourse of western marketing professionals in central and eastern Europe: Their role in the creation of a context for marketing and advertising messages'. *Discourse and Society* 9, 339–62.

Kennedy, G. (1991) 'Prooemion,' in G. Kennedy (ed. and trans.), *On Rhetoric: A theory of civic discourse*. London: Oxford University Press, pp. i–xiii.

Kitalong, K. S. (2000) '"You will" Technology, magic, and the cultural contexts of technical communication'. *Journal of Business and Technical Communication*, 14, 289–314.

Kniffin, K. (2000) 'The goods at their worst: Campus procurement in the global pillage', in G. White with F. Hauck (eds), *Campus Inc.: Corporate Power in the Ivory Tower*. Amherst, NY: Prometheus Books, pp. 36–50.

Krause, E. (1996) *Death of the Guilds: Professions, States, and the Advance of Capitalism*. New Haven, CT: Yale University Press.

Kress, G. (1996) 'Representational resources and the production of subjectivity: Questions for the theoretical development of Critical Discourse Analysis in a multicultural society,' in C. R. Caldas-Coulthard and M. Coulthard (eds), *Texts and Practices: Readings in Critical Discourse Analysis*. London: Routledge, pp. 15–31.

Kress, G. and Hodge, R. (1993) *Language as Ideology*. London: Routledge.

Krug, G. (2005) *Communication, Technology and Social Change*. London: Sage.

Labov, W. (1966) *The Social Stratification of English in New York City*. Washington DC: Center for Applied Linguistics.

Lakoff, G. and Johnson, M. (1980) *Metaphors We Live By*. Chicago: University of Chicago Press.

Lakoff, R. (1973) 'The logic of politeness; or minding your p's and q's'. *Chicago Linguistics Society*, 8, 292–305.

Lanham, R. (1993) *The Electronic Word: Democracy, Technology, and the Arts*. Chicago: University of Chicago Press.

Larson, M. (1977) *The Rise of Professionalism*. Berkeley: University of California Press.

Lassen, I. (2004) 'Ideological resources in biotechnology press releases: Patterns of theme/rheme and given/new', in L. Young and C. Harrison (eds), *Systemic Functional Linguistics and Critical Discourse Analysis: Studies in Social Change*. London: Continuum, pp. 264–79.

Latour, B. (1990) 'Drawing things together', in M. Lynch and S. Woolgar (eds), *Representation in Scientific Practice*. Cambridge, MA: MIT Press, pp. 19–68.

— (1996) *Aramis or the Love of Technology* (C. Porter, trans.). Cambridge, MA: Harvard University Press.

— (1999) *Pandora's Hope: Essays on the Reality of Science Studies*. Cambridge, MA: Harvard University Press.

— (2003) 'The promises of constructivism', in D. Ihde and E. Selinger (eds), *Chasing Technoscience: Matrix for Materiality*. Indiana: Indiana University Press, pp. 27–46.

Leech, G. (1983) *Principles of Pragmatics*. London: Longman.

Lemke, J. (1995) *Textual Politics: Discourse and Social Dynamics*. London: Taylor & Francis.

Lewis, R. G. and Smith, D. H. (1994) *Total Quality in Higher Education*. Delray Beach, FL: St Lucie Press.

Lewis, S. (1922/1950) *Babbitt*. New York: Harcourt, Brace & World.

Levinson, S. (1983) *Pragmatics*. Cambridge: Cambridge University Press.

Lyotard, J.-F. (1979) *The Postmodern Condition: A Report on Knowledge* (G. Bennington and B. Massumi, trans.). Minnesota: University of Minnesota Press.

Macey, D. (1993) *The Lives of Michel Foucault*. New York: Pantheon.

Machiavelli, N. (1961) *The Prince* (G. Bull, trans.). London: Penguin.

Martin, J. (2000) 'Close reading: Functional linguistics as a tool for critical discourse analysis,' in J. Martin and R. Veel (eds), *Researching Language in Schools and Communities, Functional Linguistics Perspectives.* London: Routledge, pp. 3–14.

Marvin, C. (1988) *When Old Technologies Were New: Thinking About Electric Communication in the Late Nineteenth Century.* Oxford: Oxford University Press.

Meister, J. (2001) 'The brave new world of corporate education'. *Chronicle of Higher Education,* 47, B10–11.

Micklethwait, J. and Wooldridge, A. (1996) *The Witch Doctors: Making Sense of the Management Gurus.* New York: Times Books.

— (2000) *A Future Perfect: The Challenge and Hidden Promise of Globalization.* New York: Crown Business.

Mills, C. W. (1951) *White Collar: The American Middle Classes.* New York: Oxford University Press.

Moore, T. E. (1997) (March) 'The corporate university: Transforming management education'. *Accounting Horizons,* 11, 77–85.

Morrell, M. and Capparell, S. (2001) *Shackleton's Way: Leadership Lessons from the Great Antarctic Explorer.* New York: Penguin Books.

Mumby, D. (ed.) (1993) *Narrative and Social Control: Critical Perspectives.* Newbury Park, CA: Sage.

Mumby, D. (1987) 'The political function of narrative in organizations'. *Communication Monographs,* 54, 113–27.

Mumby, D. and Clair, R. (1997) 'Organizational discourse', in T. van Dijk (ed.), *Discourse as Social Action. Discourse Studies: A Multidisciplinary Introduction* (Volume 2). London: Sage, pp. 181–205.

Myers, G. (1989) 'The pragmatics of politeness in scientific articles'. *Applied Linguistics,* 10, 1–35.

— (1990) *Writing Biology: Texts in the Social Construction of Scientific Knowledge.* Madison: University of Wisconsin Press.

Ni, Z. and Apple, M. (2000) 'Conquering Goliath: The free Burma coalition takes down Pepsico', in G. D. White (ed.), *Campus Inc.: Corporate Power in the Ivory Tower,* Amherst, NY: Prometheus, pp. 280–90.

Olsen, F. (2003) (11 April) 'Quid pro quo: Companies cut software donations and instead make deals with colleges.' *Chronicle of Higher Education* 49(31): A41.

Ong, W. (1988) *Orality and Literacy: The Technologizing of the Word.* London: Routledge.

Ornatowski, C. (1992) 'Between efficiency and politics: Rhetoric and ethics in technical writing'. *Technical Communication Quarterly,* 1, 91–103.

Papafragou, A. (2000) 'On speech-act modality'. *Journal of Pragmatics,* 32, 519–38.

Parenti, M. (2000) 'The myth of the liberal campus', in G. D. White (ed.) *Campus Inc.: Corporate Power in the Ivory Tower.* Amherst, NY: Prometheus, pp. 85–92.

Parker, M. (2000) *Organization Culture and Identity.* London: Sage.

Perelman, C. (1982) *The Realm of Rhetoric* (W. Kluback, trans.). Notre Dame and London: University of Notre Dame Press.

Perkins, J. and Blyler, N. (eds) (1999) *Narrative and Professional Communication.* Stamford, CT: Ablex.

Perkins, J. and Blyler, N. (1999) 'Taking a narrative turn in professional communication', in J. Perkins and N. Blyler (eds), *Narrative and Professional Communication.* Stamford, CT: Ablex, pp. 1–34.

Postman, N. (1992) *Technopoly: The Surrender of Culture to Technology.* New York: Alfred P. Knopf.

Price, D. (1965) *The Scientific Estate.* Boston: Harvard University Press.

Prince, E. F. (1981) 'Toward a taxonomy of given-new information', in P. Cole (ed.), *Radical Pragmatics.* New York: Academic Press, pp. 223–55.

Quirk, R., Greenbaum, S., Leech, G. and Svartvik, J. (1985) *A Comprehensive Grammar of the English Language.* Essex, UK: Pearson.

Rabinow, P. (1999) *French DNA: Trouble in Purgatory.* Chicago: University of Chicago Press.

— (2003) *Anthropos Today: Reflections on Modern Equipment.* Princeton: Princeton University Press.

Rankin, H. D. (1983) *Sophists, Socratics, and Cynics.* Kent: Croom Helm.

Reich, R. (1991) *The Work of Nations: Preparing Ourselves for 21st-Century Capitalism.* New York: Alfred A. Knopf.

Rheingold, H. (1999) (January) 'Look who's talking'. *Wired Magazine* 7.01, available online at: www.wired.com/wired/archive/7.01/amish.html. (27 October 2006).

Rose, M. (1990) *Lives on the Boundary.* New York: Penguin.

Savage, G. (1999) 'The process and prospects for professionalizing technical communication'. *Journal of Technical Writing and Communication,* 29, 355–81.

Schmandt-Besserat, D. (1986) 'The origins of writing: An archaeologist's perspective'. *Written Communication,* 3, 31–45.

Schmandt-Besserat, D. and Erard, M. (2007) (manuscript) 'Origins and forms of writing', in C. Bazerman (ed.), *Handbook of Writing Research.* Mahwah, NJ: Lawrence Erlbaum Associates.

Schryer, C. (1993) 'Records as genre'. *Written Communication,* 10, 200–34.

— (2000) 'Walking a fine line: Writing negative letters in an insurance company'. *Journal of Business and Technical Communication,* 14, 445–97.

Schumpter, J. (1965) *The Theory of Economic Development: An Inquiry into Profits, Capital, Credit, Interest, and the Business Cycle* (R. Opie, trans.). New York: Oxford University Press.

Scollon, R. (2001) *Mediated Discourse: The Nexus of Practice.* London: Routledge.

Simons, J. (1995) *Foucault and the Political.* New York and London: Routledge.

Simpson, L. (1995) *Technology, Time and the Conversations of Modernity.* London: Routledge.

Sperber, D. and Wilson, D. (1995) *Relevance: Communication and Cognition* (2nd edn). Oxford: Blackwell.

Spinuzzi, C. (2003) *Tracing Genres Through Organizations: A Sociocultural Approach to Information Design.* Cambridge, MA: MIT Press.

Spinuzzi, C., Hart-Davidson, W. and Zachry, M. (2004), 'Modeling knowledge work'. *Computer Writing & Research Lab Whitepaper 040505-1.* www.cwrl.utexas.edu/research/whitepapers/2004/040405-1/pdf.

Spivak, G. (1993) *Outside in the Teaching Machine.* London: Routledge.

Sullivan, W. (1831) *The Political Class Book: Intended to Instruct the Higher Classes in Schools in the Origin, Nature, and Use of Political Power.* Boston: Richardson, Lord and Holbrook.

Swales, J. and Rogers, P. (1995) 'Discourse and the project of corporate culture: The mission statement'. *Discourse and Society,* 6, 223–42.

Thompson, G. (2000) 'Unfulfilled prophecy: The evolution of corporate colleges'. *The Journal of Higher Education,* 3, 322–41.

Thompson, J. B. (1991) 'Editor's introduction', in P. Bourdieu, *Language and Symbolic Power* (J. B. Thompson, ed.). London: Polity, pp. 1–31.

Vanburen Wilkes, G. (2002) 'XML and the New Design Regime: Disputes Between Designers, Application Developers, Authors, and Readers in Changing Technological Conditions and Perceptions of Social and Professional Need'. *ACM Journal of Computer Documentation*, 26, 33–42.

Vande Kopple, W. J. (1998) 'Relative clauses in spectroscopic articles in the Physical Review, beginnings and 1980'. *Written Communication*, 15, 170–202.

— (2002) 'From the dynamic style to the synoptic style in spectroscopic articles in the Physical Review: Beginnings and 1980'. *Written Communication*, 19, 227–65.

van Dijk, T. A. (1995) 'Discourse semantics and ideology'. *Discourse and Society*, 6, 243–89.

van Dijk, T. A. (ed.) (1997) *Discourse as Social Interaction*. London: Sage.

van Dijk, T. (1998) 'Discourse and ideology (editorial)'. *Discourse and Society*, 9, pp. 307–8.

Ward, M. (2004) 'We have the power – or do we: Pronouns of power in union context', in L. Young and C. Harrison (eds), *Systemic Functional Linguistics and Critical Discourse Analysis: Studies in Social Change*. London: Continuum, pp. 280–95.

Watts, R. (2003) *Politeness*. Cambridge: Cambridge University Press.

Werkhofer, K. (1992) 'Traditional and modern views: The social constitution and the power of politeness', in R. Watts, S. Ide and K. Ehlich (eds), *Politeness in Language: Studies in History, Theory and Practice*. Berlin: Mouton de Gruyter, pp. 155–99.

Wertsch, J. (1998) *Mind as Action*. New York: Oxford University Press.

White, G. (ed.) (2000) *Campus Inc.: Corporate Power in the Ivory Tower*. Amherst, NY: Prometheus Books.

Whyte, W. (1956) *The Organization Man*. New York: Doubleday.

Winner, L. (1977) *Autonomous Technology: Technics-out-of-Control as a Theme in Political Thought*. Cambridge, MA: MIT Press.

Winsor, D. (2003) *Writing Power: Communication in an Engineering Center*. Albany, NY: SUNY Press.

Wood, L. and Kroger, R. (2000) *Doing Discourse Analysis: Methods for Studying Action in Talk and Text*. Thousand Oaks: Sage.

Young, L. and Harrison, C. (eds) (2004) *Systemic Functional Linguistics and Critical Discourse Analysis: Studies in Social Change*. London: Continuum.

Zachry, M. and Thralls, C. (eds) (2007) *Communicative Practices in Workplace and the Professions: Cultural Perspectives on the Regulation of Discourse and Organizations*. Amityville, NY: Baywood.

Zuboff, S. (1988) *In the Age of the Smart Machine: The Future of Work and Power*. New York: Basic Books.

Subject Index

This index covers all chapters except the preface. Entries in italics followed by *g* denote glossary entries on pp 185–187.

Author Index

This index covers all chapters except the preface. Only signification mentions are included.